Exploring Climate Change th Science and in Society

C000148330

Mike Hulme has been studying climate change for over thirty years and is today one of the most distinctive and recognisable voices speaking internationally about climate change in the academy, in public and in the media. The argument that he has made powerfully over the last few years is that climate change has to be understood as much as an idea situated in different cultural contexts as it is as a physical phenomenon to be studied through universal scientific practices. Climate change at its core embraces both science and society, both knowledge and culture.

Hulme's numerous academic and popular writings have explored what this perspective means for the different ways climate change is studied, narrated, argued over and acted upon. *Exploring Climate Change through Science and in Society* gathers together for the first time a collection of his most popular, prominent and controversial articles, essays, speeches, interviews and reviews dating back to the late 1980s. The fifty-five short items are grouped together in seven themes – Science, Researching, Culture, Policy, Communicating, Controversy, Futures – and within each theme are arranged chronologically to reveal changing ideas, evidence and perspectives about climate change. Each themed section is preceded with a brief introduction, drawing out the main issues examined. Three substantive unpublished new essays have been specially written for the book, including one reflecting on the legacy of Climategate.

Taken as a collection, these writings reveal the changes in scientific and public understandings of climate change since the late 1980s, as refracted through the mind and expression of one leading academic and public commentator. The collection shows the many different ways in which it is necessary to approach the idea of climate change to interpret and make sense of the divergent and discordant voices proclaiming it in the public sphere.

Mike Hulme is Professor of Climate and Culture at King's College, London, UK, having previously worked for 25 years at the University of East Anglia, UK. He established the Tyndall Centre for Climate Change Research and was its Founding Director from 2000–2007.

'I doubt that anyone on Earth can match Mike Hulme's deep understanding of both the scientific and social aspects of climate change. Yet of course what really matters, and what is so clearly on display in this volume, is the way he combines, with a sensibility that is at once rigorous and enormously generous, these two knowledge domains to provide insight and, indeed, wisdom into the true and many meanings of climate change.'

Dan Sarewitz, Arizona State University, USA

'Climate change was first an issue of climate science. But climate change is now mostly a political process, which needs recognition of its diverse cultural dimensions. Mike Hulme allows us to follow this development by presenting himself as an involved person, who has learned that climate change is not a matter of preaching the truth but of us deciding how we want to live.'

Hans von Storch, Director of Institute for Coastal Research,
Geesthacht, Germany

'Mike Hulme is reflective, scientifically precise and dispassionate. This engaging collection traces the 25-year trajectory of the writings of a fine public intellectual, as science and society become deeply entwined from climate science in the greenhouse summer of 1988 through to the era of the Anthropocene.'

Libby Robin, Australian National University, Australia and KTH
Environmental Humanities Laboratory, Sweden

'This is a fascinating collection of articles, providing a unique window on the inside world of climate change. Mike Hulme has done it all - the research, the institutions, the reflection, and the public speeches. Through his eyes, the co-evolution of climate change science and society unfolds, as it moved from the 20th Century into the 21st.'

Corinne Le Quere, Director, Tyndall Centre for
Climate Change Research, UK

'Mike Hulme's work is especially valuable because it crosses disciplines; in no field can this intellectual broadness be more essential than in the complex and bitterly contested field of climate change. His background as a physical scientist gives special weight to his insights on climate change as an evolving cultural narrative. Anyone with an interest in climate should read this book, and read it with a mind as open as Hulme's has always been.'

Mark Lynas, environmentalist and author, UK

'Here is a climatologist who has come to know that his discipline provides woefully poor impetus for political action. Whatever the prospects of averting the worst impacts of climate change, Mike Hulme is right: any proportional response must flow from deep reflection on who we are as humans, and what shapes us thus.'

Tom Crompton, Change Strategist, WWF-UK

Exploring Climate Change through Science and in Society

An anthology of Mike Hulme's essays, interviews and speeches

Mike Hulme

With a foreword by Matthew C. Nisbet

Routledge
Taylor & Francis Group
LONDON AND NEW YORK

earthscan
from Routledge

First published 2013
by Routledge
2 Park Square, Milton Park, Abingdon, Oxon OX14 4RN

Simultaneously published in the USA and Canada
by Routledge
711 Third Avenue, New York, NY 10017

Routledge is an imprint of the Taylor & Francis Group, an informa business

British Library Cataloguing in Publication Data
A catalogue record for this book is available from the British Library

Library of Congress Cataloging-in-Publication Data
Hulme, Mike, 1960-
[Works. Selections]
Exploring climate change through science and in society : an anthology
of Mike Hulme's essays, interviews and speeches / Mike Hulme.
pages cm
Includes bibliographical references and index.
1. Climatic changes. 2. Climatic changes--Social aspects.
3. Environmental economics. I. Title.
QC903.H85 2013
363.738'74--dc23
2013003206

ISBN13: 978-0-415-81162-0 (hbk)
ISBN13: 978-0-415-81163-7 (pbk)
ISBN13: 978-0-203-07007-9 (ebk)

Typeset in Goudy by
Saxon Graphics Ltd, Derby

Printed and bound by CPI Group (UK) Ltd, Croydon, CR0 4YY

Contents

List of illustrations

Figures

Tables

Foreword

Matthew C. Nisbet, Ph.D., Associate Professor and Co-Director Center for Social Media School of Communication, American University, Washington DC

Mike Hulme is the rare scientist who has made the career transition to public intellectual, influencing through his writing and commentary a global community of scholars, thinkers, journalists and engaged citizens. As an author and prolific essayist, he has infused the technical and abstract of climate science with meaning, serving as a public guide and an informed critic. In relation to policy, Hulme has courageously questioned the assumptions and approaches that we have taken for granted.

In building bridges across disciplines, he has also been a voice of authority and wisdom, encouraging us to look beyond science and economics for insight. In all of his writing, he has displayed a grace and humility uncommon to public debate over climate change, respectfully admonishing his peers for their hubris and denigration of opponents.

Public intellectuals have their greatest impact on what political scientist Amitai Etzioni calls 'communities of assumptions', the shared mental models that shape the judgements of experts, political insiders, and other journalists. These assumptions 'serve as the frameworks that influence the ways numerous specific public and private policies are received and evaluated', writes Etzioni. When shared assumptions are not available, views of complex problems like climate change become 'unsettled, cluttered with details, and lacking organizing principles and an overarching, integrating picture'.[1]

Public intellectuals also help define the fields and people who are considered authorities and quotable sources. This boundary work signals what views might be mainstream and legitimate versus what might be contrarian or out of bounds. Once assumptions and legitimate authorities are established on a problem like climate change, it becomes 'costly in terms of human mental labor to re-examine what has finally come to be taken for granted', writes Etzioni.[2]

As Hulme describes in Section One of this book, the taken-for-granted assumption over the past twenty-five years has been that increasing scientific

---

1   Etzioni, A. (2006) Are public intellectuals an endangered species? In Etzioni, A. and Bowditch, A. (eds), *Public intellectuals: an endangered species?* Rowman & Littlefield, 2006, p.9.
2   Ibid., p.7.

certainty about the causes and consequences of climate change would eventually compel countries to limit greenhouse gas emissions, even in the face of major economic costs. Once scientific certainty and agreement were reached, the central driver of change would be public education and a pricing mechanism for carbon. Within this paradigmatic policy outlook, climate change could be 'solved' mostly through the mobilisation of scientific and economic expertise alone, with few other disciplines of value or use.

Consider a striking example of these assumptions at work, as expressed by the leader of a US based foundation with more than $600 million in financing at its disposal. 'Climate change, unlike a lot of large-scale problems, is actually one that is solvable. It is also one where we know what we need to do', Hal Harvey, CEO of ClimateWorks told *The New York Times* in 2009. 'Sometimes I get accused of being too much of an engineer', Harvey said 'But sometimes with social problems, it's good to subject them to math.' Speaking specifically of ClimateWorks, Harvey declared: 'We have the best data in the world on how to prevent climate change. Everything was ranked by magnitude, location and sector. It's a systematic approach to problem solving.'[3]

Yet, not only do public intellectuals contribute to the creation of prevailing assumptions, they can also catalyse the shift to new worldviews and judgements. Political leaders and news organizations typically avoid challenging widely shared beliefs about a social problem. They instead rely on public intellectuals to lead the way, 'disturbing the canonical peace' and 'defamiliarizing the obvious' by identifying the flaws in conventional wisdom and by offering alternative renderings of a problem.[4]

In this regard, the role of the public intellectual – argued philosopher Michel Foucault – is to 'question over and over again what is postulated as self-evident, to disturb people's mental habits, the way they do and think things, to dissipate what is familiar and accepted, to reexamine rules and institutions'.[5] Conversely, in the absence of public intellectuals challenging assumptions, those working on social problems may 'be lacking in reality testing, be slower in adapting [their] policies and viewpoints to external as well as domestic changes, and be more "ideological",' warns Etzioni.[6]

In this tradition, Hulme's great contribution as a public intellectual has been to courageously defamiliarize the obvious, to confront conventional assumptions about science and policy with reality testing, and to eloquently propose alternative ways of coming to terms with climate change. Perhaps most notably in *Why We Disagree About Climate Change*, as a tool for framing a new line of thinking, he

3 Paddock, R. (2009) Climateworks is carrying out new global strategy, *The New York Times* December 5.
4 Etzioni, op. cit., p.7.
5 Foucault, M. (1990) The concern for truth, in *Foucault, politics, philosophy culture: interviews and other writings 1977–1984* Routledge, pp.255, 265.
6 Etzioni, op. cit., p.8.

popularised research from the policy sciences on the 'wicked' nature of climate change as a problem.

Like poverty, climate change is unlikely to be conventionally solved, eliminated or ended through grand policy efforts or social engineering, but rather a condition that society will struggle to do better or worse at in managing. What's needed then are not internationally mandated solutions, but a diverse menu of 'clumsy' approaches across levels of government and society. With this approach, the scope of action and conflict is reduced. Not only is there greater chance of designing effective policy, but also more hope for political agreement.

Through his work as past director of the Tyndall Centre and today as editor of *Wiley Interdisciplinary Reviews: Climate Change*, Hulme has also introduced new disciplines to our understanding of climate change, moving us beyond an exclusive reliance on science and economics (Chapters 19, 20). In doing so, he has opened doors for communication researchers like myself to be understood by scientists as valuable *social* scientist peers. More recently, by advocating for the inclusion of the humanities as part of the global research community, Hulme is once again pushing boundaries, challenging assumptions and leading the way to new understanding.

Preface and acknowledgements

There are many ways to understand the phenomenon of climate change.[7] Anthropologists may seek it by studying different human accounts of weather and agency, political scientists by studying the power relations revealed through climate negotiations between nation states. Earth system scientists are more likely to quantify and simulate the flows of energy, moisture and carbon dioxide through the planetary system. And historians will gain understanding by studying the historiography of climate change: who has written about it, when, why and how.

This book offers another way of understanding climate change: by following the written and spoken words of one person who has dedicated several decades to the professional study of the phenomenon. At its simplest this is why *Exploring Climate Change Through Science And In Society* has been compiled: a collection of essays, articles, interviews and speeches dealing with climate change which I have published or delivered over the last 25 years. It is offered as an idiosyncratic window into the changing idea of climate change.

It is rather obvious to state that how people in general understand climate change today is significantly different from how it was understood in the 1980s. Not only have physical climates around the world changed – temperatures for example are on average warmer than a generation ago and there is less ice on the planet – but the very idea of climate and its cultural and political meanings has altered even more. New social movements and business practices exist today because of climate change. New vocabularies have been invented to describe the phenomenon and new forms of artistic representation have been created.

It is perhaps less obvious to state – but still necessary – that how any one individual approaches the phenomenon of climate change has also changed over this quarter century. It is certainly true of me. And that is why this gathered and

7   The definition of the term 'climate change' is problematic – and its meaning has changed over time. In my previous book I introduced 'climate change' and 'Climate Change' to distinguish between the physical phenomenon (lower case) and the wider set of cultural discourses (upper case). This is perhaps too clumsy an approach. For the purposes of this book it is necessary to recognise that the idea of 'climate change' embraces physical (and within this both natural and human-caused change) *and* cultural dimensions of the phenomenon, and their inter-relationship.

re-published collection is of interest: it allows others to follow the evolving story of climate change as refracted through the thoughts and words of one credentialed university researcher-cum-educator-cum-public commentator.[8] For this reason all of the items reproduced here are as they originally appeared in print or as they were delivered in spoken word. What sense I made of climate change at each stage of the story – in terms of scientific understanding, political impact or cultural significance – matters if the value of this book is to be realised. This is not a heroic account of climate change science written from the perspective of 2013, even less is it a retrospective history of climate change since 1988.

Nevertheless some authorial intervention is necessary for a number of obvious reasons. The first and most substantial of these concerns which items to include in the anthology. For the most part I have not included scientific research papers, reports or book chapters. Instead I have drawn upon shorter articles which appeared in a variety of more popular printed or on-line outlets, as well as ten book reviews and a small number of speeches and interviews. A total of 54 previously written or spoken items are selected. All of these have been reproduced with permission, where needed. Five of the items were co-written with one or more authors who have also generously given their permission to reprint. The three speeches and two of the interviews included here have not been previously published. Since my career has been spent in the UK, most of the items originally appeared in public in the UK. There are 11 exceptions: seven articles were first published in the USA, two in Germany and one each in the Netherlands and Australia. Three new essays have been written specially for this book. The opening essay (Section One) offers a brief account of how climate change has evolved since 1988 as an object of scientific study and as an idea with cultural force. The closing essay (Section Nine) reflects on the wide and varied reactions to my previous book – *Why We Disagree About Climate Change* – drawing upon published and unpublished reviews and correspondence. There is also a new previously unpublished essay (Chapter 50) evaluating the lessons and legacies of Climategate.

Related to the selection of items is the manner of their ordering. Rather than present the 54 items simply in chronological order, they are grouped into seven sections dealing with the following themes: Science, Researching, Culture, Policy, Communicating, Controversy and Futures. The items in each section appear in chronological order. Each section has a brief introduction which summarises the content and makes connections between the chosen items, including cross-references to chapters in other sections which may share the same or similar context.

The second key editorial decision is when to date the start of this collection. The earliest items included here date from 1988, a year chosen for two reasons. This was the year when I first entered into employment at the University of East Anglia, in

---

8   A brief biography of my career was offered in the Preface to my 2009 book *Why We Disagree About Climate Change*.

the university's Climatic Research Unit based in the School of Environmental Sciences, following a period of four years' employment at the University of Salford. The year 1988 is also significant for other reasons. It was the year of 'the greenhouse summer' in the United States and on 27 September 1988 Margaret Thatcher made a speech at the Royal Society in London, the first by a major political leader dealing with the greenhouse effect. That the book will be published almost exactly a quarter of a century after Thatcher's speech is a tidy landmark, particularly as it is appearing in the year of her death.

The third editorial intervention concerns the editing and formatting of the selected items. With just three exceptions, all of the items are reproduced here as they originally appeared in public, either in print or on-line (minus some copyrighted photographs or other illustrations). The three exceptions are Chapters 6, 8 and 19 which have been somewhat shortened in order to leave out extraneous material which was pertinent for their original context, but not for this one. Spelling and grammatical errors have been corrected in all re-printed items, references updated and formats and linguistic conventions standardised, but no opinions have been altered. In some cases where specific reports and other contemporary items are referred to, these references have been added to improve traceability. In the case of some on-line essays web-hyperlinks have been removed and replaced with footnotes. All originally referenced material has been collated into one master Bibliography which appears at the end of the book.

What are the limitations to this collection? As I have made clear it is not a consolidated history of climate change from 1988 to 2013, nor is it a climate change reader: a balanced selection of the most important work on climate change published over the last quarter century. It is simply the selected work of *one* analyst and commentator working within a UK university context, but who has spoken in over 40 countries and been involved with many of the scientific, policy and cultural institutions, initiatives and events dealing with climate change over this period: for example, the Intergovernmental Panel on Climate Change, UK and EU climate policy forums, the Tyndall Centre for Climate Change Research, public and media engagement, student education, book festivals, Climategate, businesses and non-governmental organisations.

Acknowledgements

I would like to thank Asher Minns of the Tyndall Centre for Climate Change Research who several years ago first suggested the idea for a volume such as this. I also would like to thank the many colleagues and students here at the University of East Anglia (in the Climatic Research Unit, the Tyndall Centre, the Schools of Environmental Sciences and International Development, the Science, Society and Sustainability (3S) Group, the Faculty of Humanities), and around the world, who over the 25 years of work represented here have acted as provocateurs, mentors and critics. They are too innumerable to mention individually, but they represent the web of human relationships within which my work has been conducted and without which there would be no story to tell.

I am grateful to Louisa Earls in the sustainability division of Routledge for enthusiastically adopting the idea of the anthology when I first proposed it to her and to the production team at Routledge, especially Helena Herd, Jennifer Birtill and Helen Bell. Three anonymous reviewers for Routledge encouraged the project along and made some useful suggestions as did Max Boykoff, Reiner Grundman, Mathis Hempel, Martin Mahony and Sam Randalls who read and made helpful comments on the three new essays included here. I would also like to thank my co-authors – Elaine Barrow, Suraje Dessai, Rob Lempert, Martin Parry, Roger Pielke Jr. and Jerry Ravetz – for allowing me to reproduce our jointly written items. Sarah Clarke of the School of Environmental Sciences at UEA re-typed a number of the earlier articles in this collection and Bill Johncocks produced, as always, a wonderfully complete and professional index across this sprawling mass of material. Permissions to reproduce all previously published material have been secured for all items subject to copyright. Specific acknowledgements are included as appropriate at the beginning of each chapter. The cartoons introducing each section, and also the cover illustration, were designed by Stephen Collins.

Articles by chronology

Articles by category

Books reviewed

On-line articles

Speeches

Interviews

New essays

Section one

The public life of climate change

The first twenty-five years

Since 1988, science, politics, culture and ethics have exerted changing influences on the idea of climate change. The ways in which climate change is deployed in public life have diversified and proliferated. This essay describes some of these changes that have taken place over the last 25 years.

January 2013

Into the light

On the evening of Tuesday 27 September 1988 in central London, at the annual dinner of the Royal Society, the British Prime Minister Margaret Thatcher delivered a speech which had wide-ranging ramifications for the politics of climate change. It is not often that a serving British Prime Minister delivers a speech to the Royal Society, but then there have not been many Prime Ministers who have been trained scientists. Indeed Margaret Thatcher is the only one. She studied chemistry at Oxford University and then worked briefly as a research chemist before entering politics. She was elected a Fellow of the Royal Society in 1983 at the beginning of her second term of office. The central subject of Thatcher's 20 minute speech in the Fishmonger's Hall in the City of London was the environment and in particular the greenhouse effect and climatic change. Whilst praising the enterprise of science – and UK science in particular – Thatcher warned of 'a global heat trap which could lead to climatic instability' and referred to the possibility that 'we have unwittingly begun a massive experiment with the system of this planet itself'.[9]

Of course the notion that humans might be altering climate on a planetary scale had a long scientific pedigree, extending back through Roger Revelle and Guy Callendar to Svente Arrhenius and John Tyndall in the nineteenth century

---

9 www.margaretthatcher.org/document/107346 (accessed 14 December 2012).

(see Chapter 14). And already by 1988 there was a vibrant and growing band of climate scientists scattered across a handful of countries who saw their primary task as increasing understanding of the 'massive experiment' Margaret Thatcher was referring to. By drawing out 'the wider implications for policy' of these scientific insights, the British Prime Minister was very deliberately claiming that climate change and its human dimensions was a matter for political attention. Mrs Thatcher was the first senior world leader to turn human interference with the climate system into a national and international policy issue of the first importance. The headlines in the UK print media the following day revealed the significance of the speech: 'Cautious greeting to Thatcher's greening' (*The Guardian*) and 'Thatcher gives support to war on pollution' (*The Times*). Even the leading British science journal *Nature* asked the question in their editorial ten days later 'Has Mrs Thatcher turned green?'[10] The journal answered this question 'no', but it welcomed the additional attention that Thatcher's speech would inevitably bring to this 'potentially serious problem'.

Earlier in the summer of 1988, the greenhouse effect had also made a star appearance in American politics and public life. On 23 June, at a US Senate hearing orchestrated by Senator Tim Wirth, American climate scientist Jim Hansen testified that it was time to 'stop waffling so much and say that the evidence is pretty strong the greenhouse effect is here'.[11] With the American mid-West in drought and the outside air temperature in Washington DC that day over 100 degrees Fahrenheit, the US media widely reported Hansen's claims. The greenhouse effect was now a public celebrity, marked by *Time Magazine* nominating Earth as 'Planet of the Year' for 1988 in their cover story for 2 January 1989. As Steve Schneider remarked in his journal *Climatic Change*: 'It was with some pleasure that I observed this viewpoint [of humans causing unprecedented climatic change] passing from the ivy-covered halls of academe and the concrete and glass of government offices into the popular consciousness' (Schneider, 1988: 113).

Exploring Climate Change through Science and in Society is being published 25 years after Mrs Thatcher's speech in the Fishmonger's Hall and in this opening essay I offer a very brief history of climate change over the ensuing quarter century. The argument I develop in the paragraphs below is simple. During these 25 years not only has the world's climate changed, but the ways we think about and represent the relationships between climate and human agency have also changed. And it is these latter changes that are perhaps even more significant than the measureable changes in physical climate. This essay therefore provides context for my changing views on climate change which are reproduced in the following chapters. But of course it does so with the benefit of perspective and reflection which the passing of time alone offers. This is not something which the content of the essays, interviews and speeches reproduced

10 Editorial, *Nature* 335, 479 (6 October 1988).
11 'Global warming has begun, expert tells Senate' *New York Times*, 24 June 1988.

here benefitted from; they are reproduced as they originally appeared at the time.

Climate change: 'the plan'

As we have seen, the greenhouse effect first became an object of public conversation in the Western world in the late 1980s. In his study of early US media coverage of the phenomenon, journalism analyst Lee Wilkins showed the order of magnitude increase in print coverage given to the story which took place between the years 1987 and 1990 (Wilkins, 1993). Furthermore, he showed how in these early days the greenhouse story in the United States was framed in three ways: as a fact revealed by scientific progress, as a 'loss' of a pristine and unspoiled climate and as a problem which scientists and government officials were dealing with. Two important background events helped shape these dominant frames which were beginning to emerge. One was the Montreal Protocol on Substances that Deplete the Ozone Layer, signed in 1987 and ratified in 1989. The other was the demise and subsequent collapse in 1991 of the Soviet Union and its various spheres of influence. The former event suggested that scientific evidence could catalyse a global regulatory framework for atmospheric pollutants; the latter that a new era of American dominance in a unipolar world was beginning.

These two events were central in establishing what Dan Sarewitz has called 'the plan' (Sarewitz, 2011) – the dominant assumptions which both frame the problem of climate change and offer the solution. For Sarewitz these assumptions were, first, that knowledge about climate change – its causes and potential impacts – would drive forward both a common understanding of the problem and also a widespread commitment to take action. The second assumption of 'the plan' was that the action to which all would be committed would be a reduction in greenhouse gas emissions, thereby reducing the risks for society of climate change. Following the 'greenhouse summer' of 1988, institutional support for these two assumptions began to take shape. The IPCC published its first assessment report in 1990, a report which was adopted by governments to provide the knowledge base for guiding negotiations towards a multi-lateral global agreement on climate change. And the UN Framework Convention on Climate Change (UNFCCC) was signed at the Earth Summit in Rio de Janeiro in June 1992 and ratified two years later.

Following the publication of the IPCC's second assessment report in 1996, the UNFCCC negotiating process resulted in the Kyoto Protocol being signed in 1997. The Kyoto Protocol adopted a targets-and-timetables approach to mitigation, whereby signatory nations agreed to reduce their emissions of greenhouse gases by specified amounts from a 1990 baseline by the period 2008–2012. So-called 'flexible mechanisms' were created which included carbon-trading and the award of emissions-credits for carbon offsetting investments in non-signatory poorer nations. The EU's emissions trading scheme (ETS) became the flagship policy instrument to deliver 'the plan', using market mechanisms to 'put a price on carbon' and hence drive down emissions across Europe and incentivise investment into lower carbon energy technologies.

The IPCC's third assessment report was published in 2001 and the launch of the EU ETS and the ratification of Kyoto Protocol followed in 2005. 'The plan' seemed on track. But the withdrawal of the USA from the Kyoto Protocol in 2001 was a major setback and so new entrepreneurial ventures emerged to bolster 'the plan' and to persuade sceptical or disinterested citizens and businesses to buy-in. Thus climate change came to Hollywood in 2004 with the blockbuster disaster movie *The Day After Tomorrow*, which dramatised the consequences for the United States of (super-) abrupt climate change. Box-office takings were high, but viewing the movie was as likely to induce apathy as activism amongst members of the public (Lowe *et al.*, 2006). More significant was Al Gore's 2006 movie *An Inconvenient Truth*. Although this film also gained mass audiences, unlike *The Day After Tomorrow* it generated a trans-national movement for promoting its message of 'Small actions by billions of people will help to solve our climate crisis'.[12]

'The plan' was further boosted in late 2006 with the publication by the British government of the Stern Review of the economics of climate change (see Chapter 31). This brought together climate science, economics and policy in a powerful way and using the language of investment showed why 'the plan' could work if the right levers were pulled and pushed. The following year the IPCC published its fourth assessment report. As if to sanctify its contribution to 'the plan' (and to world peace?), the IPCC was subsequently awarded – jointly with Al Gore – the 2007 Nobel Peace Prize (see Chapter 53). This fourth assessment report offered the strongest claims yet about attributing 'most' global warming to human activities and it gave a further impetus to the stuttering UNFCCC negotiating process. By now this was in danger of stalling with a cheerleading EU, a reluctant USA, the newly emerged superpower of China and a coalition of smaller and poorer nations all pulling in noticeably different directions. But at the 13th negotiating session of the Conference of the Parties (COP13) to the UNFCCC in Bali in December 2007, a road-map for achieving a new legally binding climate treaty was triumphantly agreed. The road from Bali was to pass through Poznan (COP14) before culminating in Copenhagen (COP15) in December 2009.

According to the above truncated account, the twin processes of the IPCC and the UNFCCC were central in shaping and driving forward a political response to climate change. A globally forged scientific consensus was bound rhetorically to a multi-lateral UN convention and negotiating process. This combination of authorities and institutions broadly replicated the pattern of diplomacy that was successfully implemented in the late 1980s around stratospheric ozone. Climate change therefore co-produced a new form of global science and an international political order in the manner described by Clark Miller (Miller, 2007). But 'the plan' crashed in the winter of 2009/10. Climategate and 'glaciergate' (see Section 7) undermined the power and virtue of climate science and COP15

---

12 http://www.takepart.com/an-inconvenient-truth (accessed 17 December 2012).

at Copenhagen revealed the dominance of national interests over any putative enlightened global political order. At 'possibly the most important meeting mankind has ever had'[13] the hoped-for fair, ambitious and binding (FAB) deal evaporated in front of 13,500 participants.

Contesting 'the plan'

The above account is of course a caricature, but in essence it captures the dominant framework within which climate change science and policy operated over the two decades from 1988 to 2009. It is the framework advocated by the IPCC, the UNFCCC, the EU and other powerful international and Western institutions. However at every stage in the development of this narrative – on every page of the story – challenges have been mounted. One of the earliest of these was to the credibility of the IPCC's first assessment report in 1990. Even as the IPCC reports were being finalised over the summer of that year, the prominent campaigning organisation Greenpeace was publishing its own account of global warming and what should be done about it. The book was edited by Greenpeace's director of science, Jeremy Leggett, and in it he accused the IPCC of 'failing in its responsibilities' in refusing to listen to warnings from climate scientists. As Leggett boldly describes, 'This then [*Global Warming: the Greenpeace Report*] is the book which says what the IPCC – in order to be consonant with the warnings of its own scientists – *should* have said about how we must respond to the greenhouse threat' (Leggett, 1990: 8, emphasis added).

Two things are interesting here. First is how Greenpeace saw the 'correct' response to global warming emerging directly and unambiguously from the scientific evidence. There was only one way to interpret and act on this evidence and this was to reduce greenhouse gas emissions through an immediate radical programme of energy efficiency, energy innovation and technology transfer funded through a tax on armaments. Second, is to note that Greenpeace's challenge to the IPCC came long before – 18 years in fact – another organisation with a very different ideology – the Heartland Institute – offered its own challenge to the IPCC.[14] But Greenpeace was not alone in its criticisms of the IPCC in 1990. The journal *Nature* was also critical. In an editorial published in November 1990 to mark the occasion of the Second World Climate Conference in Geneva and a few months after the publication of the IPCC's report, *Nature* called for 'a more persuasive framework of argument than IPCC has provided', criticised the IPCC for its 'failure to discuss dissenting opinions' and lamented that the Working

---

13 Thomas Schultz-Jagow, Stop Climate Chaos campaign http://www.stopclimatechaos. org/tags/COP15?page=12 (accessed 17 December 2012).
14 In 2008, in response to the IPCC Fourth Assessment Report published the previous year, the Heartland Institute published its own assessment report, the Nongovernmental International Panel on Climate Change (NIPCC) (Singer, 2008).

Group 2 and 3 reports dealing with impacts and responses were, respectively, 'inadequate' and 'naive'.[15]

Even while Leggett was commissioning, editing and publishing the Greenpeace Report, the world's political climate was changing more rapidly than its physical climate. The Berlin Wall came down in November 1989, Nelson Mandela was released from prison in February 1990 and between 1989 and 1991 the Soviet Union and its hegemony dissolved. Francis Fukuyama's *The End of History and the Last Man* appeared in 1992 to offer a polemic in favour of the ultimate and eternal triumph of Western liberal democracy. As if to show-off this Cold War victory, the Earth Summit which took place in Rio de Janeiro in June 1992 put in place new global environmental governance arrangements. Over 150 leaders of sovereign nations turned up to sign a number of environmental treaties, including the UNFCCC which had as its goal the avoidance of dangerous climate change through the stabilisation of greenhouse gas concentrations in the atmosphere. Yet Rio also showed that unlike stratospheric ozone depletion, dealing with climate change was going to need to engage with the politics of international development (see Chapter 6). Troubling and disputed questions of equity, justice and human rights could not be separated from the challenges of limiting climate change. And such questions were not those on which science could adjudicate, however powerful and purposeful a consensus the IPCC might muster. The supremacy that science had claimed in revealing, shaping and solving the issue of climate change was misplaced.

This particular challenge to 'the plan' was clearly brought home early in 1991 when two Indian analysts Anil Agarwal and Sunita Narain confronted the early framing of human-induced climate change as an environmental problem. In their influential pamphlet *Global Warming in an Unequal World* (Agarwal and Narain, 1991), these authors argued that questions of equity and justice lay at the very heart of climate change. Scientists may treat each ton of carbon dioxide identically in terms of radiation physics, but each ton of carbon dioxide carried very different moral valencies depending from where, by whom and for what purpose it was emitted. Luxury emissions needed to be distinguished from survival emissions. The seeds of the climate justice movement were sown.

A different sort of challenge to the dominant framing of climate change also became institutionalised during the 1990s. The new environmentalism which emerged in the 1960s had always antagonised a variety of corporate and industrial interests, ever since Rachel Carson's book *Silent Spring* appeared in 1962. And so, 30 years later, a new coalition of interests opposed to severe or premature constraints on fossil fuel emissions formed in the United States and elsewhere. The Global Climate Coalition (GCC), comprising large corporations with interests in fossil fuels, was founded in 1989 and provided a powerful counter-narrative to 'the plan'.

---

15 Editorial 'Next steps on global warming', *Nature* 348, 181–182 (15 November 1990).

Attacking the credibility of the IPCC and the claims of climate scientists was a favoured tactic. Thus in a 1997 article in *New Scientist* magazine titled 'The greenhouse wars', self-declared climate contrarian Pat Michaels railed against the IPCC: 'They can't go on forever tinkering with their models, trying to make them fit reality. Ever heard of Ockham's razor? It says the simple explanation is usually the best. Apply that in this case and you conclude that the climate just is not as sensitive to the greenhouse effect as they predicted.'[16] The GCC eventually disbanded in 2002 as different member companies sought to express their own self-interest in different ways, but a few years later the free-market think-tank The Heartland Institute took up the mantle of the GCC and became a powerful institutional voice against the wisdom of 'the plan'. In this they were supported by growing numbers of individual voices who began to make effective use of new social media to 'audit' the claims of climate science and policy.

By the beginning of the twenty-first century realisations of a changing world order were inescapable. The new millennium saw the centre of world economic activity shifting away from the Atlantic to the Pacific and the tropics. The new century was being claimed by some as belonging to China and new coalitions of large economies such as BRIC (Brazil, Russia, India, China) and the G5 (Russia, India, China, Mexico, South Africa) showed the emergence of a multipolar world. The Kyoto Protocol coming into force in 2005 did nothing to stem the rise in global emissions of greenhouse gases. It was to take a world financial crisis to do this, and even then only for one year (2009). Efforts focused increasingly on how to engage legitimately these newly emerging economies in emissions reductions. The UN's multi-lateral framework established in 1992, with a rigid divide between developed and developing nations dating back to the world of the 1980s, was looking increasingly anachronistic (see Chapter 34).

For this reason significant voices began to question whether mitigation strategies really were the most appropriate or achievable ones for dealing with climate change. Adapting to climate change – a view which had been marginalised in the 1990s because it did not fit 'the plan' (see Chapter 27) – now became a more central part of the discourse. The Marrakesh Accords signed at COP7 in December 2001 used the language of 'adapting to the adverse impacts of climate change' and also established the possibility of adaptation finance for poorer nations. Yet this move towards adaptation again brought climate change face-to-face with the volatile questions of justice and development. And later in the decade, as emissions from BRIC countries soared, a third strategy for dealing with climate change began to be championed. This was climate remediation (or geoengineering), a third response strategy beyond mitigation and adaptation. This, too, did not fit 'the plan' and its curious collection of speculative technologies

16 Pearce, F. (1997) The greenhouse wars *New Scientist* (19 July), London. http:// www.newscientist.com/article/mg15520915.100-greenhouse-wars.html (accessed 17 December 2012).

for controlling global temperature or atmospheric carbon dioxide concentrations quickly triggered vigorous arguments – mostly in North America and Western Europe – over whether even to *research*, let alone deploy, such technologies was legitimate (see Chapter 35).

Re-framing climate change

These challenges to 'the plan' emerged from the mobilisation of diverse political, cultural, economic and corporate interests. They illustrate how climate change has come to mean different things to different people in different places at different times. But it is also important to realise that science itself has been re-framing climate change during this quarter century. By this I don't mean the simple notion that climate or Earth System science has been 'filling the gaps in knowledge' or 'narrowing the uncertainties' of climate predictions, which is how the IPCC in its first and second assessment reports in the 1990s described its future tasks. This would suggest science progresses in a linear manner, always sure of its eventual object of understanding, only being thwarted in realising its goal of gaining a faithful account of reality 'as it really is' by incomplete theory, lack of observations or inadequate simulation tools. Rather, what we have seen is that the philosophy and practice of climate science – through its framing of key questions, its forms of representation, its use of metaphor, its public communication – keep changing the form and locus of the 'climate reality' being studied.

There are many examples of this. Over the 25 years since 1988 changing scientific conceptions of the climate system have given much greater prominence to the role of non-linear thresholds and feedbacks. Whereas most of the graphs of future climate change in the IPCC in 1990 were linear, the recognition and simulation of complexity in Earth System models increasingly allows for tipping points, hysteresis loops and dynamical thresholds. Indeed, the idea of 'tipping points' shows the power of a metaphor to transform our conceptions of physical reality. Although used in the social and medical sciences for several decades, this metaphor was only introduced into climate science in 2005, but has since become a dominant framing device in both science and policy advocacy (Russill and Nyssa 2009), not always with predictable outcomes.

Another example of how changes in scientific philosophy and practice have altered the imaginative and practical implications of climate change is in descriptions of the future. Whereas in the earlier part of our period future climates were described in terms of scenarios (possibilities with unknown likelihood), more recent quantitative descriptions of future climate change attach likelihoods or probabilities to the outcomes (see Chapter 9). These different forms of representation lead to very different psychologies of risk and to different philosophies of decision-making. Indeed, the wider argument is that changes in how climate modelling uncertainties are represented re-defines how the idea of climate change is deemed relevant for adaptation strategy and decision-making (see Chapter 13). Thus recent years have seen a shift away from a predict-and-adapt paradigm to one of robust-decision-making and decision-support (Weaver *et al.*, 2013).

This paradigm change has itself been part of a wider re-structuring of the relationship between climate science and society. The dominant focus of climate science to develop and improve climate predictions on the time-scale of decades-to-centuries is giving way to a new focus on weekly-to-seasonal-to-decadal prediction. This change in modelling strategy has implications for policy discourse by changing the relative weights given to different causal explanations of climate change and variability. Whereas long-lived greenhouse gases dominate centennial-scale climate, a subtle mix of natural factors and short-lived radiative forcing agents are more important on decadal time-scales and for 'weather'. This shifting ground in climate science carries implications for institutional design and policy formation. The idea of 'climate services' is gaining ground through initiatives such as The Global Framework for Climate Services. Rather than being driven by questions concerning detection and attribution of human influences and long-term prediction of climate, climate science is being directed to providing climate information that can 'facilitate climate-smart decisions that reduce the impact of climate-related hazards and increase the benefit from benign climate conditions' (Hewitt *et al.*, 2012: 831).

And the whole field of climate attribution studies is also changing shape. Rather than asking the question which dominated the 1990s and 2000s, 'have we detected human influence on the climate system?', climate scientists are now investigating a rather different question: 'what is the probability that this extreme weather event was caused by [greenhouse gases/soot emissions/dust particles]?' Here too changes in scientific framing have the potential to re-configure political interests around climate change. The investigation of weather event attribution opens up the possibility of a new policy discourse around international legal liability for weather loss and damage, although there remain many unresolved difficulties in this manoeuvre (Hulme *et al.*, 2011).

There are many ways of framing the political, economic and cultural challenges of a changing climate and the above examples show how the changing paradigms and emphases of scientific research carry influence in this struggle for framing. The 'handles' offered by science for understanding the phenomenon of climate change keep changing.

Changing representations

In 1988 few serious commentators really believed that the politics of climate change would be anything other than tortuous. Yet the assumption has remained through this period that human-induced climate change is an important, urgent and discrete problem which at least in principle lends itself to policy solutions. Optimism has waxed and waned, but the belief has been maintained that at least some forms of policy intervention will yield tangible public benefits – whether in terms of reductions in climate-changing emissions (from some assumed baseline) or else through reduced exposures to climate risks. Yes, the climatic side effects of large-scale combustion of fossil fuels were an unforeseen and undesirable outcome of Western and then global industrialisation. But putting this causal chain into reverse – arresting some of these unwanted side-effects – was believed to be in the reach of an intelligent, purposeful and ingenious humanity. This presumption must now be questioned. Maybe the climate system cannot be managed by humans.

And this brief survey of climate change over 25 years has shown at least two reasons why. First, there is no 'plan', no self-evidently correct way of framing and tackling the phenomenon of climate change which will over-ride different legitimate interests and force convergence of political action (see Chapter 42). Second, climate science keeps on generating different forms of knowledge about climate – different handles on climate change – which are suggestive of different forms of political and institutional response to climate change. Or put more generally, science as a form of creative inquiry into the physical world co-evolves with the physical phenomena it is seeking to understand. Taken together these two lessons suggest other ways of engaging with the idea of climate change, not as a discrete environmental phenomenon to prevent, control or manage, but as a forceful idea which carries creative potential.

Innovations in the representation of climate change in recent years suggest that various actors and institutions have recognised this potential. By this I mean that some scholars, artists and social entrepreneurs are discounting the scientific, environmental and managerial dimensions of 'the plan', whilst promoting the argument that the idea of climate change carries much fruitful imaginative significance (see Chapter 40). These interventions are less interested in halting global warming at two degrees than they are in using climate change to give form and voice to different expressions about what it means to be human. There are different ways of portraying this re-imagination of climate change: as a cultural turn, as a crisis in representation or by suggesting that climate change has become a synecdoche – climate change as a term which 'stands-in' for something else. But there is little doubt that compared to 25 years ago there is a greater emotional, aesthetic and spiritual engagement with the idea of climate change (see Chapter 20).

This cultural turn – let's settle for that – is evident from a casual survey of the thousands of books about climate change which have been published over this quarter century. Back in 1988 the books being published on climate change topics typically dealt with science, geography or environment themes.[17] Twelve years later around 2000 there were many more books dealing with the economic, political and

---

17 Out of the books published this year here are a typical selection of titles: *Physically-based modelling and simulation of climate and climatic change* (Schlesinger, 1988); *The impact of climate and man on land transformation in Central Sudan* (Ahlcrona, 1988); *The impact of climatic variations on agriculture* (Parry et al., 1988).

management dimensions of climate change.[18] But in recent years not only has there been an order of magnitude increase in the overall number of books about climate change being published, we see all aspects of human life now analysed or represented in relation to climate change: gender, violence, literature, security, architecture, the imagination, football, tourism, spirituality, ethics and so on.[19] All of human life is now lived out not just in the presence of a physically changing climate/planet, but in the new discursive and cultural spaces which have been created by the idea of climate change (Endfield and Morris, 2012). It is as though all human practices and disputes now can be – have to be? – expressed through the medium of climate change. So photography, cartoons, poetry, music, literature, theatre, dance, religious practice, architecture, educational curricula, and so on, are now expressed through this medium. And political disputes about landscape aesthetics, child-rearing, trade tariffs, theology, patents, extreme weather, justice, taxation, even democracy itself, find themselves inescapably caught up in the argumentative spaces and linguistic expressions of climate change. Climate change has become a new medium through which human life is now lived.

The marks of climate change

It is with this changing idea of climate change and its multiple forms of representation that my own professional career has engaged and evolved. Throughout these 25 years I have been employed by the University of East Anglia. I was studying climate change in 1988 and I am still studying climate change in 2013. And yet the object of my study has changed, as too have many of my convictions, assumptions and methods. *Exploring Climate Change through Science and in Society* – by bringing together a collection of my written and spoken work on climate change over this period – offers some insight in how this journey has left its mark on one individual. In 1988 I was concerned about the variability of African dryland climates (see Chapter 1), I worked mostly with climatological observations and I believed that research into climate change would change the world (for the better). In 2013 I am concerned with the multiple meanings of climate change to people in different settings, I work mostly with texts, images and speech and I believe that the idea of climate change has the power to change individuals (for the better).

---

18 Out of the books published this year here is a typical selection: *Warming the world: economic models of global warming* (Nordhaus and Boyer, 2000); *Climate for change: non-state actors and the global politics of the greenhouse* (Newell, 2000); *Sustainable forests management and global climate change* (Dore and Guevara, 2000).

19 A handful of examples: *Climate change for football fans: a matter of life and death* (Atkins, 2010); *Gender and climate change: an introduction* (Dankelman, 2010); *Between God and green: how evangelicals are cultivating a middle ground on climate change* (Wilkinson, 2012); *Climate change and national security: a country-level analysis* (Moran, 2011); *Climate change and future justice: precaution, compensation and triage* (MacKinnon, 2012); *English and climate change* (Leal Filho and Manalos, 2012).

Section two

Science

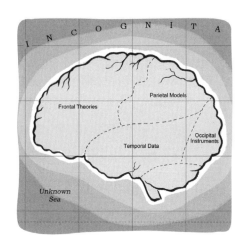

Introduction

The miscellany of articles selected in this section on science represents how I have commented upon and interpreted aspects of climate science for assorted public audiences over the last 25 years. These articles are not a selection of my scientific research papers which, of course, had much more specialised audiences in mind.

The selection starts in 1988, the year when greenhouse gas-induced climate change first emerged into the full glare of public and policy interest (see Section One). In the spring of that year I had started writing for the London-based *Guardian* newspaper, pioneering a new monthly retrospective UK climate summary (which I kept going for nearly 13 years) and also writing short articles on topics dealing with climate change which I felt were of some public interest. The first five selections in this section (Chapters 1 to 5) are drawn from this latter series and offer brief perspectives on the following topics: drought in the African Sahel, changing frequencies of Atlantic hurricanes, a multi-century perspective on British winters, nuclear autumn and the performance of climate simulation models. It is noticeable how themes of scientific uncertainty, caution and skepticism are evident, long before these terms became emotionally charged in the context of the later cultural politics of climate change.

Research for my PhD in the early 1980s had got me thinking about the relationships between climate and development, especially in dryland Africa, and as human-induced climate change gained in policy significance in the late 1980s I wrote an article on this topic for a rather obscure journal, *Science, Technology and Development*. A shortened version of this article from 1990 is reproduced here (Chapter 6) which, two years before the UN Framework Convention on Climate Change was signed at Rio, emphasises some of the reasons why I thought climate change might be important for developing countries.

The early 1990s also witnessed the publication of the first (controversial) estimates of global temperature trends derived from satellites. The next item (Chapter 7) is a review for *New Scientist* magazine in 1991 of William Burroughs' book *Watching the World's Weather* in which he explains how satellite technologies had changed atmospheric science. Whilst recognising these benefits, I was also cautious against his claims that such technologies would of themselves reduce the damage and death associated with weather-related hazards. I argued that the human dimensions of such disasters would still need dealing with through institutional, social and political change. It was success or failure here that would determine the benefits of improved weather monitoring and climate prediction using satellites.

Until September 2000 I worked in the Climatic Research Unit (CRU) in the School of Environmental Sciences at the University of East Anglia and in 1997 CRU marked the 25th anniversary of its founding by Professor Hubert Lamb. By then, Hubert was in his last year of life and one of the projects we undertook to commemorate this anniversary was publishing an edited book in his honour: *The Climates of the British Isles: present, past and future*. The opening section of the Introduction to this book – jointly written with my co-editor Elaine Barrow – is included here as Chapter 8, since it presages a number of themes about my understanding of climate-society relationships which were to mature over the next 15 years. In particular, I warned against the dangers of climate determinism infecting both climate change science and rhetoric, a warning I was to repeat a decade later (Chapters 23 and 24).

By the late 1990s moves were afoot in climate science and modelling to develop more probabilistic predictions of future climate change. The next item (Chapter 9) was written for the in-house magazine of the Environment Agency, the UK's regulatory body charged with – amongst other things – managing climate risks, especially flooding. This article offers a short explanation of the sources of uncertainty in climate predictions and my endorsement of the move towards representing these in probabilistic terms. At the turn of the millennium I was also interested in the question of extreme weather event attribution, a topic which has generated new methodological advances in climate science only in the last few years. The devastating flooding in Mozambique in March 2000 gave me an opportunity, in a commentary written for the environment section of *The Guardian* newspaper (Chapter 10), to make the observation that all weather events are now, to a greater or lesser extent, human-influenced. My point was that one should not think of two different types of weather: natural weather and human-caused weather and try to distinguish one from the other. No, the functioning of the planetary system is now to some extent irrevocably altered by human activities, just as, for example, are ecosystems and our body chemistry. All the manifestations of that system – whether routine or extreme – are therefore now different from what they would otherwise be.

In February 2005, the UK Government commissioned an international conference on 'Avoiding dangerous climate change', convened by the Met Office in Exeter. It was part of the British Prime Minister's climate change diplomacy ahead of that summer's G8 meeting in Gleneagles, Scotland. The main papers of that conference were published in book form a year later, which in August 2006 I reviewed for the *Times Higher Education Supplement* (Chapter 11). As well as pointing out the politically motivated origins of the Exeter conference, I also argued that the apocalyptic rhetoric emerging from it about climate change catastrophes would be detrimental to making political headway. The limits of science for driving forward political change had been reached.

Unusual weather events always present an opportunity to write about climate change and in 2008 the UK experienced a rather wet and gloomy summer. I used this moment to re-visit and critique some of my own earlier work on climate scenarios and climate adaptation. In a commentary for *BBC News On-line* (Chapter 12), I suggested that too much emphasis of long-term multi-decadal climate scenarios might lead to mal-adaptation by under-emphasising the inherent short-term interannual variability of regional climate. This was a theme I returned to a year later with three colleagues in our editorial for the house magazine of the American Geophysical Union, *EOS: Earth Observation Science* (Chapter 13). We argued for a robust approach to adaptation decisions which de-emphasised the reliance of long-term climate predictions.

May 2009 marked the 150th anniversary of the original experiments performed by John Tyndall in London which established empirically the radiation absorptive properties of greenhouse gases such as water vapour and carbon dioxide. To commemorate this occasion, I wrote an essay capturing Tyndall's work and thoughts from that time, drawing upon his published papers and unpublished diaries. Written for the popular UK magazine *Weather*, this essay (Chapter 14) captures some of Tyndall's interpretations of his own experimental work, set in the context of London's scientific and political elite with whom he mixed.

1 Sahel awaits the rain

22 April 1988

An article written for *The Guardian*'s climate column[20]

The rainy season is due to start over most of the semi-arid zone of the Sahel in Africa next month. Farmers, especially, are anxious. The last 20 monsoons in Sudan, in the eastern Sahel, have brought rainfall well short of the average for the twentieth century. Twenty years is a long time for Sahelians – nearly half their average life expectancy. It is also long enough for climatologists to define a new climatic environment, much less favourable than the wetter decades of the colonial era.

There are at least three different levels of explanation. At a local level, more intensive land use in the Sahel (which also covers Mauritania, Mali, Niger, Senegal, Burkina Faso and Chad) had led to a reduction of surface vegetation. This reduces the amount of energy reflected from the land and enables larger loads of dust to enter the atmosphere. These consequences are thought likely to reduce rainfall. At a global level, rainfall reduction in the Sahel is linked with forces in the Pacific Basin (El Niño events in which a huge area of the eastern Pacific warms by several degrees) and the possible shift of whole belts of the atmosphere's circulation towards the equator.

Between these local and global extremes there is an intermediate level of explanation which has yielded the most promising results in terms of both understanding and forecasting. The temperature difference between the oceans of the northern and southern hemispheres is the key. With a relatively warm south Atlantic, the moist monsoon airflow into the Sahel is weakened; with a relatively cool south Atlantic the monsoon is enhanced.

What is now clear is that for the last 20 years or so the south Atlantic has been relatively warm. This has therefore undermined the rainfall regime of the Sahel. What remains less clear is the future warming of the southern oceans.

20 Hulme, M. (1988) Sahel awaits the rain *The Guardian* 22 April.

2   Sea heat fuels hurricanes

15 September 1988

An article written for *The Guardian*'s climate column[21]

Hurricanes are officially defined as tropical storms during which wind speeds reach more than 34 m/s (75 mph) for one minute or more. Such storms are given personal names, like Gilbert, but anything less is described merely as a tropical cyclone.

There are six main regions in the world where hurricanes develop and these are restricted to tropical areas. Generally, ocean temperatures of at least 26°C are required to generate sufficient evaporation to subsequently provide the latent heat release necessary to maintain the intense energy of the cyclone.

As oceans reach their maximum temperature in the autumn, this is the hurricane season – Gilbert, which hit the Caribbean this week, is a spectacular example. The north-west Pacific hurricane region is the most productive with an average of about 25 occurring each year. But the most destructive is usually in the Bay of Bengal and in the western Atlantic. The western Atlantic's annual average of eight hurricanes during this century includes periods of high and lower frequencies: between 1900 and 1930 for example the average was about six and this rose to ten between 1930 and 1960. Very few hurricanes have passed through the Caribbean during recent years. In 1979 hurricanes David and Frederick caused extensive damage to the Leeward Islands and last September Emily reached 57 m/s as it passed over Haiti.

The enormous potential for destruction has led to efforts to predict their intensity, frequency and tracks. A computer-based experiment in the USA recently suggested that rising ocean temperatures in the worldwide greenhouse effect could increase hurricane intensity by 30 to 40 per cent by the middle of next century. There is less certainty about whether frequencies and tracks would also change.

21 Hulme, M. (1988) Sea heat fuels hurricanes *The Guardian* 15 September.

3   Cold facts about winters

27 October 1988

An article written for *The Guardian*'s
climate column[22]

In Britain, we are fortunate in possessing some very early regular air temperature measurements. Several of these records were used by the late Professor Gordon Manley to reconstruct a continuous series of mean, monthly temperatures, back to January, 1659. This series is now updated monthly and is referred to as Manley's Central England Temperature since the stations he used are representative of the English midlands. It contains 330 years of nearly precise data and is the longest climate series of its kind in the world. We can use this series to examine trends in British winter weather over the last three centuries.

If we define winter as December through February and date it by the year of the January, then the coldest winter was 1684 with a mean temperature of −1.2°C. The next coldest winter in the series (−0.3°C) occurred in 1963 and will be remembered by many for the prolonged snow cover between Christmas and March. The coldest single month was January 1795, which recorded −3.1°C, far colder than anything we have encountered in recent years. At the more pleasant extreme, the warmest winter occurred in 1869 and had a mean temperature of 6.8°C. The warmest winter of recent years was in 1975 (6.5°C).

Between 1659 and 1988, the average winter temperature has been 3.6°C and 90 per cent of winters have recorded temperatures between 1.7 and 5.3°C.

A question which a lot of people ask is, 'Have our winters been getting warmer or colder?' The answer to this, of course, depends on the length of the period we look at. Since the end of the seventeenth century, there has been a slight warming of British winters of just under 1°C. This is largely because of the colder winters of 1680–1820 during the end of the Little Ice Age.

Most of this winter warming occurred in the century between 1830 and 1930. If we look just at our present century, there has been no overall trend for warming or cooling. The occasional severe winter has occurred (1917, 1947, 1963) and there has been the occasional mild winter (1935, 1975 and 1988). In very recent years, although we have had some very cold individual months (February 1986 was −0.9°C and December 1981 −0.3°C), no winter has been very far below the overall average of 3.6°C.

---

22 Hulme, M. (1988) Cold facts about winters *The Guardian* 27 October.

4   Nuclear autumn danger

25 November 1988

An article written for *The Guardian's* climate column[23]

Last month was the fifth anniversary of the publication of the results from the first global model which simulated the effects of a nuclear war on the earth's climate. The model suggested that the huge volume of smoke injected into the atmosphere by the blasts, and more importantly by the subsequent surface fires, would reduce surface temperatures by up to 20°C over continental areas for prolonged periods. Lesser cooling effects would spread into the southern hemisphere. This catastrophic by-product of a nuclear war was termed 'nuclear winter'.

Figure 1 'Mushrooms: the heralds of autumn'

23 Hulme, M. (1988) Nuclear autumn danger *The Guardian* 25 November.

In the past five years considerably more has been discovered about these likely post-nuclear climates. We can identify four main things: A nuclear war is likely to have a far bigger impact on the Earth's climate if it is fought in the northern hemisphere summer rather than in winter since a much greater amount of solar radiation would be intercepted by the smoke. More sophisticated computer models of the Earth's climate now suggest that the resulting temperature decrease would at most be 10°C during the summer and a few degrees during winter. This has translated 'nuclear winter' into 'nuclear autumn (Figure 1)'. Even a nuclear autumn, however, would have chronic effects on agriculture, vegetation and animal life. By altering the length of growing seasons and changing temperatures and rainfall, severe damage would be done to a range of the more sensitive plant and animal species. And, most recently, the substantial warming (+20°C to 50°C) which is predicated for the stratosphere at the outer margins of the atmosphere, could well upset the chemistry that protects the ozone layer.

Depletions in the ozone of up to 50 per cent could occur, not just over the Poles where current concern is focussed, but over large sections of the whole northern hemisphere. Although the original severity of the nuclear winter predicted by the 1983 model has been scaled down, the clear likelihood of a major global climatic catastrophe following exchanges in a nuclear war remains with us. The explosion of just a small proportion of existing warheads would trigger the atmospheric feedbacks. The ensuing chaos and damage to our environment is clearly a factor in arguments for or against the nuclear deterrent. It could be disastrous unless we make sure that, as the Prime Minister said recently in a different but relevant context, such a 'massive experiment with the system of the planet itself' is never carried out.

5 Generalists in hot pursuit

17 February 1989

An article written for *The Guardian*'s climate column[24]

To estimate climatic changes caused by increasing concentrations of carbon dioxide and other greenhouse gases, scientists rely on theoretical models of global climate called General Circulation Models (GCMs). These models can currently only represent average conditions in the atmosphere for areas about 300 km square and about 2 km deep. GCMs therefore generate climate that do not completely agree with observations of the present climate state, do not produce perfect estimates to climatic conditions over the next few decades, and do not fully agree with each other.

At present, while the handful of powerful GCMs around the world broadly agree on the large-scale features of future annual and seasonal climate change, they generate quite different results for individual regions and for shorter time scales such as days and months. As the size of the region gets smaller, the agreement between the different models gets weaker. Table 1 shows results from the five leading GCMs in the world. When the concentration of greenhouse gases is doubled, the models agree that summer will become wetter by between 7 and 15 per cent for the world as a whole. But for the UK there is no agreement about what would happen, not even whether it would get wetter or drier. Temperature estimates show similar, although smaller, discrepancies.

Scientists must remain sceptical about the ability of these GCMs to protect future climatic changes for regions such as the UK or even northwest Europe. But this, of course, is exactly the scale for which other environmental scientists (agriculturalists, hydrologists, foresters) and governments are beginning to demand accurate projections of daily and monthly scale climate changes to enable them to develop appropriate management strategies.

Climatologists still only have a broad-brush picture of how world climates will change over the next 50 years. To begin to fill in the finer details will require a sustained research effort, including improved computing capabilities, more sophisticated climate models, and more extensive international scientific collaboration. This will be an exciting, but costly, task.

24 Hulme, M. (1989) Generalists in hot pursuit *The Guardian* 17 February.

Table 1 Change in average summer temperature and precipitation – for the world and for the UK – for a doubling of carbon dioxide concentration, obtained from five different global climate models

	Average summer temperature change (°C)		Average summer precipitation change (per cent)	
	World	UK	World	UK
Model 1	+3.3	+3.3	+7	+10
Model 2	+4.1	+5.9	+9	−17
Model 3	+4.3	+3.9	+11	−12
Model 4	+3.0	+3.1	+8	−3
Model 5	+4.2	+5.0	+15	+38

6 Global warming in the twenty-first century: an issue for Less Developed Countries

April 1990

An article published in *Science, Technology and Development*[25]

The decade of the 1980s brought an increased ability for the human species to monitor the health of its only known environment ... Earth. This improved ability has been underpinned by three key developments: an increased understanding of the interactions between different biogeophysical components of the global systems; the introduction of routine monitoring from satellites of a wide range of geophysical parameters; and the increased global penetration of telecommunication technologies. The latter have enabled the rapid and effective dissemination around the world of environmental data collected with national and international research projects.

The decade of the 1980s has also brought with it the highest global-mean surface air temperatures which have been recorded since the commencement of widespread instrumental meteorological measurements in the nineteenth century, and possibly since the middle of the last interglacial 120,000 years ago. Globally, the 1980s have been about 0.2°C warmer than the 1950s and about 0.5°C warmer than the 1900s, with 1988 being the single warmest year (Jones *et al.*, 1988a). The cause of this warming is not yet known unequivocally. The strongest candidate is that it has resulted from increasing global concentrations of atmospheric trace gases, mainly carbon dioxide, methane, nitrous oxide, chlorofluorocarbons and low-level ozone. These gases are collectively known as greenhouse gases because they are transparent to incoming short-wave radiation. Their net effect is to trap heat in the lower layers of the atmosphere thus causing surface air temperatures to rise. This phenomenon is known as the greenhouse effect and threatens to alter the global climatic environment to conditions never before experienced by the human species.

This article summarises current evidence for, and future projections of, global warming, and argues that this likely global climate change is a major issue for Less Developed Countries (LDCs).

---

25 Hulme, M. (1990) Global warming in the twenty-first century: an issue for Less Developed Countries *Science, Technology and Development* 8, 3–21. This article has been edited and shortened.

Historical change in global-mean climate

Changes in global-mean temperature can be estimated by combining together meteorological observations from the land and oceans areas of the Earth, the bulk of which were originally obtained as routine data for weather forecasting purposes. The marine data are particularly important, because they represent some 70 per cent of the Earth's surface area (although coverage of this area is incomplete, simply because it is limited to the regions where ships travel). Many hundreds of millions of observations go into producing area-average figures spanning the last 100 years or so. Before they can be used, these data must be examined critically for inhomogeneities (that is, for variations arising from non-climatic sources, such as changes in station location, observing times, measurements practise and effects such as urban warming).[26]

Jones *et al.* (1986) have produced the most comprehensive analysis to date. They show that the near-surface air temperature averaged over the globe has increased by about 0.5°C since the late nineteenth century (see Table 2). The recent global-scale warming has not been a continuous upward trend. Nor has the warming been spatially homogeneous, and trends have varied substantially from region to region. Over the past 40 years, for example, much of the Atlantic and Western Europe has undergone a slight cooling (Jones *et al.*, 1988b). This cooling appears to be inconsistent with the greenhouse hypothesis. Although the cooling is undoubtedly real, it cannot be taken as evidence that there is no greenhouse effect, nor even that the magnitude of the greenhouse effect is small. Rather, it is a graphic illustration of the extent and magnitude of natural climatic variability, the noise against which the greenhouse signal must be detected, and upon which future greenhouse warming will be superimposed.

Table 2 Historical changes (c.1890–1990) in global-mean surface temperature, terrestrial precipitation and global-mean sea-level. Hemispheric contrasts are provided for temperature and precipitation. Precipitation units are relative

	Surface Temperature		Terrestrial Precipitation	Sea-level
Global	+0.5°C ± 0.1°C		+0.04	+12.5cm ± 2.5cm
Northern Hemisphere	+0.4°C	Northern Tropics	−0.09	
Southern Hemisphere	+0.7°C	Southern Tropics	+0.12	

26 An assessment of the magnitude of such biases and a discussion of their significance for the global-mean temperature time series can be found in Jones *et al.* (1991). There remains an unresolved uncertainty in the consistency of late nineteenth century ocean temperatures, which may affect the scale of global warming since 1860 by ±0.1°C.

Parallel changes in global-mean sea-level have also occurred. Compilations of tide gauge data from over 300 sites worldwide show a rise of about 12cm since the late nineteenth century (Barnett, 1984). As with global-mean temperature, such a rise has been neither a continuous upward trend nor spatially homogeneous. Changes in global-mean precipitation are more difficult to assess for two reasons. First, long time series of precipitation measurements are restricted to land based sites, thus immediately between 60 per cent and 70 per cent of the globe is excluded from any historical analysis. Second, the greater spatial and temporal variability of precipitation make reliable estimates of global-mean precipitation in mm units hard to calculate. The most comprehensive analysis published to date is presented in Bradley *et al.* (1987) and Diaz *et al.* (1989). Some results suggest a marked difference in precipitation trends between the northern (decreasing) and southern (increasing) tropics. Again, the high noise/signal ratio of precipitation ensures that no definitive greenhouse signal may be discerned from these trends, although at least in the case of the African Sahel a plausible mechanism for linking decreased precipitation and warmer oceans has been proposed (Parker *et al.*, 1988).

Projected changes in global-mean climate

The magnitude of the projected change in global-mean climate in the twenty-first century depends on four key factors: the rate of increase in the emission of greenhouse gases; the estimated sensitivity of the global climate system to greenhouse-gas-induced thermal forcing; the time required for the climate system to approach equilibrium (that is, the magnitude of the thermal inertia effect of the oceans); and the magnitude of any non-anthropogenic climate forcing, such as caused by volcanic eruptions or solar variability. Uncertainty exists on all four counts.

Global warming: an issue for LDCs

Climate affects human activities everywhere. It has been argued that owing to the different structure of societies the total impact of climate goes much deeper in LDCs than in developed countries (Sah, 1979). The agricultural base of many LDC economies is taken to imply that climate influences the availability of goods and capital, prices, employment, growth prospects and the well-being of people to a far greater extent than in the developed world. A frequently used example of this viewpoint is the observation that the 1984–85 drought in northeast Africa resulted in about 100,000 deaths, whereas the 1976 drought in the UK caused only water rationing. If we accept this argument then it might be extended in the following way. If global climate changes during the next 100 years at the rate and to the extent discussed above, then the effects of such change will be felt more keenly in LDCs than in developed nations. If agriculture production is greatly affected under current climate variability, how much more will it be affected by future large-scale climatic change?

There are a number of reasons to hesitate from reaching such a conclusion. First, temperature changes are generally projected as being substantially lower in the tropics than in mid-to-high latitudes (Schlesinger and Mitchell, 1987). Second, interannual variability of precipitation is already high in the tropics and subtropics and might well remain substantially more important in determining annual yields than possible changes in mean precipitation of ±10 per cent or ±20 per cent. This idea is elaborated in Parry *et al.* (1988). The third point follows on from the previous one. Economic and agricultural activity in mid-to-high latitudes is now more finely tuned to a given climatic regime than is such activity in the tropics. Although traditional cropping patterns in the tropics are not as widespread as previously, where such traditional agriculture is still practised it is capable of operating under and responding more quickly to a wider range of climatic regimes. Finally, it may be argued that climatic change will take a lower political priority in LDCs than in developed countries. A substantial number of LDCs are confronted with management problems of far greater potential impact than global warming. These would include civil war (for example, Sudan, Sri Lanka, Mozambique), epidemics (for example AIDS in Uganda, Zaire), water resource provision for rapidly expanding populations (as in Egypt), or coastal inundation due to recurrent cyclones (as in Bangladesh). In view of such reasoning, why should LDCs place particular urgency on addressing the potential disruption to their societies posed by global warming? Three reasons will be suggested.

Future contribution of LDCs to atmospheric greenhouse gas concentrations

It was previously stated that the developed nations are currently the major contributors to global carbon dioxide emissions on both an absolute and a per capita basis. The average American citizen for example releases about five tons of carbon per year compared to about 0.3 tons by the average African. Although these rations will probably narrow over the next few decades, there will still be considerable differences between LDCs and developed countries in their per capita generation of carbon dioxide (and probably other greenhouse gases). These give the LDCs a powerful bargaining position in international negotiations aimed at stabilising global carbon emissions (the other greenhouse gases are more problematic because sources are harder to identify and emissions rates harder to quantify).

A recent report by Michael Grubb of the Royal Institute of International Affairs in London has explored what options exist for obtaining a global agreement on carbon emissions (Grubb, 1989). His conclusion is that the most feasible system would involve a global market in which carbon emissions quotas could be leased between countries. Each nation would have the right to emit up to the global average per capita carbon output (currently about two tons per person per year). Nations which were to emit at levels above this could do so, but only by leasing additional per capita permits from nations which were to emit at levels lower than their entitlement. A leasing market would hence be created where, predominantly, LDCs (because of their low per capita carbon output) would be leasing permits to developed countries (which because of the extent of industrialisation and high energy consumption would want to keep emitting carbon at high levels). The lease price would be determined by the newly created market, and the currency would be monetary, but tied specifically to development or environmental enhancement projects in the recipient nation. Thus, under this system, LDCs would have a new commodity to trade on the global market which would result in substantial revenue to be spent on high priority environment and development programmes.

This proposed system of stabilising carbon emissions has many points in its favour and is being discussed in Working Group III of the Inter-Governmental Panel on Climate Change (IPCC) due to report to the United Nations in December 1990. From the LDC point of view it provides an additional incentive to become fully integrated into the international processes whereby global warming preventative strategy will be determined. If not, the agenda and resulting agreements will continue to be dominated by developed world interests. This new diplomatic power that the global warming issue is creating for LDCs is being recognised by some in the developing world (for example, Cheng, 1989), but it is vital that a widespread caucus of LDC politicians and policy-makers is established. The recent global climate conferences in New Delhi (February 1989), Cairo (December 1989) and Nairobi (May 1990) are important components of awareness building in LDCs.

Regionally specific impacts of global warming

While the general impacts of global warming in LDCs might be of smaller magnitude than those resulting from 'natural' climate variability, there are a number of severe impacts to be anticipated in certain regions of the developing world. A rise in global-mean sea-level will cause substantial problems for low-lying coastal states such as Egypt, Bangladesh, the Maldive Islands, and some other Pacific Ocean and Indian Ocean states. Most major cities in both developed countries and LDCs will be protected from sea-level rise, but at great expense. In LDCs, however, sea-level rise will be most severely felt by exposed coastal populations and by agricultural developments in deltaic areas. India, Bangladesh and Egypt are likely to be especially vulnerable because the highly populated deltas of these countries are already subject to storm surges. Since 1960, India and Bangladesh have been struck by at least eight tropical cyclones that each killed more than 10,000 people (Bardach, 1987). Recent estimates suggest that a climatically induced one metre sea-level rise would inundate arable land in Egypt and Bangladesh presently occupied by 8 and 10 million people respectively (Broadus *et al.*, 1986).

Availability of, and access to, shared water resources are two other concerns for LDCs which are likely to be exacerbated by regional climate changes following global warming. Of the 13 international river basins shared by five or more countries, ten are shared mostly by LDCs (Gleick, 1989). High on this list are the

Niger, Nile and Zambezi basins. Even in the absence of substantial climate change, pressure on the water resources of these basins has increased due to population growth, industrial water demand, and development. The hydrological changes anticipated following global warming, although poorly specified as yet, will greatly complicate the operation and management of the existing systems of aqueducts, reservoirs and dams which were designed for past climatic conditions not future ones. In basins where water resources are widely shared, these physical effects will have important political implications for LDCs. A preliminary study of the sensitivity of precipitation over the Nile Basin to global warming suggests that although White Nile discharge may increase slightly, any change in Blue Nile discharge would not be enough to offset increased evaporation losses following a rise in surface air temperatures (Hulme, 1992). The net result would be to alter the seasonality of Main Nile discharge and reduce Sudan's hydroelectric power generating capacity on the Blue Nile. The costs and benefits of precipitation changes within shared river basins will be felt unequally and will exacerbate existing political tensions.

Indigenous research on climate monitoring an impact assessment

In 1979 Raaj Sah identified four priorities in climate-related activities for LDCs (Sah, 1979). These were, strengthening the capability for applied climatological work including the collection and maintenance of climatological data; improving the ability for short-term weather prediction in the tropics; participating in multilateral research into tropical climates co-ordinated by the World Meteorological Organisation (WMO); and undertaking exploratory research on weather modification techniques. The challenge of global climate warming reiterates the first and third of these priorities for the scientific and meteorological institutes of LDCs.

Global climate research during the 1980s has become substantially more international in its orientation, and at the same time more dependent on the efficient collection and dissemination of global climate observations. Such observations are largely the responsibility of the national meteorological agencies; the WMO exists to enable and facilitate the subsequent exchange of data. LDCs have a key role to play in the international research effort which will further scientific understanding of global warming. The effective monitoring of regional climates and subsequent maintenance of good quality datasets will remain a key element of international climate research over the next decade. Although many national meteorological centres in LDCs do give high priority to this aspect of their activities (as in Zimbabwe, India, Malaysia), there are also those where an injection of urgency and priority is needed. This need was reflected in one of the conference conclusions of the 'International Conference on Global Warming and Climate Change: Perspectives from Developing Countries' held in New Delhi in February 1989 (Woods Hole Research Centre, 1989). It was recommended that 'a national Climate Monitoring, Research and Management Board be established in countries where such an organisation does not really exist ... such a Board should monitor on a continual basis the state of the atmosphere' (p.19).

The other area where global warming provides an impetus for research in LDCs is climate impact assessment. There is a wealth of environmental scientists working in LDCs who can bring local expertise and insight to some of the regional-scale questions of how ecosystems will respond to global climate change. A European initiative (Landscape-Ecological Impact of Climate Change, LICC) has recently been launched from the Netherlands which is aimed at creating a base network of environmental scientists specifically concerned with climate change impacts (LICC, 1990). Similar such regional groupings in the developing world would be advantageous. A good recent example of illuminating work performed largely by research institutes in tropical countries is the reassessment of the significance of tropical grasslands in the global carbon cycle.[27] The research, sponsored by UNEP and performed in Kenya, Mexico and Thailand, has shown that the carbon-storing capacity of tropical grasslands is substantially greater than previously thought. This has major implications for the global modelling of carbon fluxes and future projections of carbon dioxide concentrations.

The priority areas of such climate impact for LDCs should be in the area of river basin management (as discussed above), crop water use efficiency and yield sensitivity to enhanced atmospheric carbon dioxide, impact of climate change on pest fertility and migration, and perhaps most important the key role of tropical forests, both as a regulator of regional (global) climate and as an ecosystem threatened by global climate change.

Conclusions

The progressive increase in the anthropogenic contamination of the global atmosphere through emissions of a range of greenhouse gases has a high likelihood of perturbing the global climate system to an extent whereby the nature of all regional climates will be substantially altered. Although such regional changes are as yet hard to identify unambiguously, there is a wide consensus that by the early decades of next century, global climate will be measurably different from that experienced over the last few millennia of human existence. One inevitable consequence of future global warming will be a rise in global-mean sea-level. These projected changes will occur unevenly and their impacts will be felt unequally between nations. Although LDCs are as yet a minor contributor to this ill-advised experiment with the Earth's climate, and although they currently face a range of pressing economic, political and resource management problems not shared with the developed world, it is imperative that LDCs do not sideline themselves in the growing number of diplomatic, economic and scientific negotiations and programmes.

27 Reported in *New Scientist*, 6 January 1990, p.29.

The unequal contribution to increasing greenhouse gas concentrations between LDCs and developed nations transform a hitherto impotent inequality (per capita carbon, and other greenhouse gas, emissions) into a new and potentially powerful trading commodity which LDCs should exploit to the full. The value of this commodity, however, will devalue with time as current differentials in per capita greenhouse gas emissions between LDCs and developed nations diminish. Although subject to some potentially devastating impacts from current 'natural' climate variability (for example, El Niño impacts in South America, cyclones in Bangladesh, and drought in Africa), LDCs would be ill-advised to consider future global warming either as merely a concern of the privileged industrial nations, or as a secondary concern which may be dealt with as and when the full impacts become clear.

This article has outlined three reasons why this conclusion is reached, although there are undoubtedly more. First, LDCs need to bargain hard for a deal on the stabilisation of global atmospheric carbon concentrations; they are currently in a favourable position and should be wary of squandering this by pursuing development routes which demand high greenhouse gas emissions. Second, certain regional impacts of global climate change will be disastrous for some LDCs (for example, the sea-level rise in areas of deltaic agriculture). Third, global climate and climate impact research can give large stimulations to research institutions in LDCs.

7 How good is technology's weather eye?

25 May 1991

A review of: *'Watching the world's weather'* by William Burroughs, written for *New Scientist* magazine[28]

It was perhaps unfortunate that I read this book just after the inundation of large coastal areas of Bangladesh and the announcement that the subsequent death toll was probably more than 200,000. On the second page of W. J. Burrough's book *Watching the World's Weather*, the claim is made that the advent of weather satellites has both dramatically improved forecasts of the movements of storms, such as the severe cyclone of 1970 which led to about 250,000 deaths in Bangladesh, and sharply reduced the loss of life.

The first part of this claim is probably indisputable (although improvement does not imply accuracy); the second is definitely worth a lively discussion. The recent devastation in Bangladesh was caused by a tropical cyclone which was quite well forecast with the aid of satellite imagery. To save lives, however, good forecasts of extreme weather require good responses.

Despite the many benefits that satellite-based environmental monitoring has introduced to society, the improvement of social, institutional and political decision-making and response structures is not one of them. The on-going haemorrhage of life in northeast Africa is further evidence that good quality and reliable satellite-derived information (in this case concerning rainfall and crop performance) is only the first, and relatively easy, step toward 'sharply reducing the loss of life'.

I was also uneasy for other reasons about the faith in human technological achievement that underlies this optimistic, attractive and well-illustrated book. Burroughs explains how we can watch the world's weather from satellites. The assertion is made, with such frequency that it almost appears as the main justification for meteorological satellites, that this is important for understanding climate and climate change. But there is a distinction between weather and climate that needs to be made before this assertion can be evaluated.

Weather can be seen but climate cannot. We experience weather directly through hail storms, floods or hurricanes; climate is a statistical construct that

28 Hulme, M. (1991) *How good is technology's weather eye?* Book review of 'Watching the world's weather' by W. J. Burroughs *New Scientist*, 25 May, p.48.

has proved of immense value for planning purposes. Belief in the invisible is hard, which explains why climatology was a relatively late disciplinary arrival on the horizon of human knowledge. Belief that the invisible changes, however, is even harder – thus explaining why climate change developed as a serious offshoot of climatology only within the past two or three decades.

The geological evidence of the Ice Ages unearthed in the eighteenth and early nineteenth centuries led to the belief in climate change on very long timescales. That climate changes on contemporary timescales has only been realised, at least in most circles, in more recent years through the systematic analysis of early documentary and instrumental evidence. This difference between weather and climate commonly leads to confusion when we try to explain contemporary climate change to a popular audience – whether the requirement to understand this distinction as part of the geography National Curriculum at Level 5 (9 to 10 year olds) will reduce this confusion remains to be seen.

The direct evidence we all have of changes in weather (a windy January, a Great Storm or a record hot summer) is often mistakenly used to infer climate change. Since climate is a statistical construct, a change in climate can be inferred only from a change in statistics, which by definition are gathered over a long period of time. Here lies the limitation of satellite 'watching' for climate studies – the longest homogenous satellite time series of any component of the global climate system is no more than 20 years long and in many cases much shorter.

When one compares the efficacy of satellite versus ground-based observation of the world's weather for climate change purposes exactly the same problems emerge: massive data accumulation of which only a fraction is analysed; instrumental drift or failure leading to inhomogeneity in a time series; change in observation site due to orbit decay or platform replacement again leading to inhomogeneity in a time series; and observational error. Here the famous example is the rejection by the automated data screening procedure of the measurements made by the Total Ozone Mapping Spectrometer (TOMS) on Nimbus 7 in the mid-1980s of the substantial and increasing depletion of stratospheric ozone over Antarctica.

All these problems also affect climate change monitoring using conventional ground-based instrumentation, although there is one sense in which ground-based data are more appropriate for climate change detection. The average lifetime of a satellite instrument, by design, is usually only a few years – few if any meteorological satellites have operated for more than a decade – compared with a potential lifetime of two or three decades for conventional meteorological instruments. Replacement costs, too, are vastly more favourable for ground-based monitoring ensuring that, with care, long-term homogeneity of meteorological time series can be maintained (of critical importance when small climate changes are to be detected).

This is not to decry the key role that meteorological satellites have in furthering our understanding of the global climate system – they increasingly provide extensive baseline data of new, and not so new, meteorological variables for weather forecast and climate models. But it is to warn against a false vision of the future in which all the needs of weather and climate science are met through the 'eye in the sky'. It is one thing to watch the world's weather and to forecast it reliably; it is another matter to reduce vulnerability to weather-related hazards and to detect global climate change.

My final evidence for the technocentric advocacy I detect at times in this book is the author's title for the second chapter 'The global weather machine'. A machine implies a human creator. Whether one is a theistic or atheistic evolutionist, a Gaian or Creationist, the global climate system most certainly is not made by humans.

8 Introducing climate change

8 May 1997

The introductory chapter, co-authored
with Elaine Barrow, of the edited book
*'Climates of the British Isles: present, past
and future'*, published to mark the
25th anniversary of the Climatic
Research Unit[29]

> The more things change, the more they are the same.
>
> Alphonse Karr

Changing views of climate

With hindsight, the twentieth century view of climate was dominated by the perspective that climate is constant. Climate was effectively stationary for the purposes of human decision-making and only varied in any significant way over geological time. This view of climate was partly conditioned by the dominance of the developments in weather forecasting which took place during the first half of the century. The excitement of scientific discoveries which led to improvements in the understanding and prediction of the weather of the next day marginalised work which was more concerned with variations in climate over decades and centuries. The adoption of 'normal' periods by the fledgling International Meteorological Organisation reinforced this rather static view of climate. Weather statistics collected over thirty or thirty-five years were thought of as adequate to define the climate of a region, statistics which could then be safely used in future design and planning applications. Climate change was largely irrelevant.

This was in contrast to much thinking in the nineteenth century. The evidence of glaciation discovered during the early decades of that century and the emergence of evolutionary ideas were more consistent with a dynamic view of nature and of climate than one in which all things remained constant. Concern amongst colonialist conservationists about deforestation in the tropics causing

---

29 Hulme, M. and Barrow, E.M. (1997) Introducing climate change pp.1–7 in *Climates of the British Isles: present, past and future* (eds) Hulme, M. and Barrow, E.M., Routledge, London, UK. This chapter has been edited to remove extraneous editorial material related to the organisation of the book.

climate change – in this case loss of rainfall (Grove, 1994) – was also consistent with this view. Thus a major English national newspaper could observe in 1818 that,

> a prospect far more gloomy than the mere loss of wine had begun to present itself by the increased chilliness of our summer months. It is too well known that there was not sufficient warmth in the summer of 1816 to ripen the grain; and it is generally thought that if the ten or twelve days of hot weather at the end of June last had not occurred, most of the corn must have perished. The warm and settled appearance of the weather at this early period of the season, leads us to hope that an agreeable change is about to take place in our planet; and that we shall not, as for many past years, have to deplore the deficiency of solar heat which is so necessary to ripen the productions of the earth.[30]

Only during the last quarter of the twentieth century – from the 1970s onwards – has this more dynamic view of climate been rediscovered. The twentieth century view that climate is constant was first seriously challenged by a few pioneering scholars, of whom Hubert Lamb must rank in this country as perhaps the most important. They were followed by the growing body of climate scientists and, by the time of our present decade, by an increasing constituency of decision-makers. A strong sense of history was characteristic of those originally challenging the twentieth century orthodoxy. Indeed, Hubert Lamb and others were almost as much historians as they were climatologists. More recently, events in the climate system itself reinforced the challenge and have now led to a re-writing of the orthodoxy.

The prospect of significant global climate change induced by human pollution of the atmosphere has acted as a powerful agent in consolidating the revisionist view of climate as non-stationary. This process of re-thinking has been underpinned by the twin developments of more abundant global climate observations and rapid increases in computer modelling capability. It is now possible to describe truly global changes in climate using observational data and to explore future changes in climate using credible climate models. The changed attitude towards climate has also been institutionalised in recent years. In 1988, for example, the World Meteorological Organisation and the United Nations Environment Programme established an Intergovernmental Panel on Climate Change to assess the evidence for the enhanced greenhouse effect, or so-called 'global warming'. This Panel continues to produce reports for the world community on the prospect of climate change (IPCC, 1996a) and they have also considered the consequences of global climate change for individuals, ecosystems and nation states (IPCC, 1996b). The concern about changing global climate was sufficient to yield a United Nations Framework Convention on Climate Change. This

30 *The Observer* newspaper, London, 18 June 1818.

Convention was signed by 155 nations at Rio de Janeiro in June 1992 and subsequently came into force in March 1994. The British and Irish governments ratified the Convention in December 1993 and April 1994, respectively, and both diplomatic delegations have played their part in the on-going negotiations to establish a legally binding climate protocol.

These developments have taken place against a background of a warming climate. Since the 1970s, both the British Isles and the world have warmed by about 0.3°C. The reality of this warming, and the prospect of accelerated warming over the next few decades, has focused more attention on the interactions and interdependencies of climate and society than was hitherto the case. Thus the United Kingdom Government through its Department of the Environment commissioned national reviews in 1991 and again in 1996 of the potential impacts of climate change for the country (CCIRG, 1991; 1996). This type of national review of the importance of climate change is required of many countries under the Framework Convention and is a mode of reporting that has been adopted around the world.

There is a danger that this recent political concern about climate change and its impacts bestows on climate an unwarranted importance as an agent that shapes our lives. Such thinking has led, perhaps rather curiously, to a return in some quarters to a variant of the climatic determinism prevalent at the start of the century. Determinism is a reductionist philosophy that sees events and behaviour as controlled by a very limited set of physical factors. Ellsworth Huntington, the Yale geographer, is the most well-known proponent of such a role for climate. He argued in 1915 that, 'The climate of many countries seems to be one of the great reasons why idleness, dishonesty, immorality, stupidity, and weakness of will prevail' (Huntington, 1915: 294).

Although not always as strident or doctrinaire as Huntington, the importance of the climatic influence on our lives has been stressed by numerous thinkers, starting with the Ancient Greeks and their supposedly uninhabitable, torrid and frigid 'climata'. The influence of climate has also been interpreted psychologically. In the middle of this century, for example, Gordon Manley stated that, 'Appreciation of the British climate depends largely on temperament. That it has not been conducive to idleness has been reflected in the characteristics of the people' (Manley, 1952: 15) and, more recently, Richard Beck argues that, 'the historical record is highly suggestive … that a mild climate in mid-latitudes helps to foster a tolerant society or that an extreme climate may predispose people towards intolerance' (Beck, 1993: 63–64). These psychological interpretations of climatic determinism may seem hard to defend. Nevertheless, the prospect of global warming, and the study of the impacts of such climate change, introduces a new variant to the climatic determinists' repertoire of arguments.

Many studies of the possible impact of future climate change seem, implicitly, to elevate climate to being the major factor that will influence future human activity and welfare. Thus the conventional climate change impact study would attempt to simulate the effect of climate change by, say, 2050 on a particular aspect of the environment, say cropping patterns or forest distribution. Little attention is usually paid to whether or not climate is the main driving factor behind observed changes in such distributions. Even if it is recognised explicitly that other factors are involved (e.g., changes in technology, consumer behaviour, work and leisure patterns), these are so unpredictable that climate often retains the appearance of being the main controlling factor. Climate change determinism thus re-appears. Some studies have shown, however, that factors such as the future of the Common Agricultural Policy of the European Union will have a much larger impact on the future British landscape than climate change (Parry *et al.*, 1996). And it only takes a simple thought experiment to realise that other considerations, too, will swamp the effects of climate change on future human and animal welfare. For example, civil conflict, technological and demographic change and global epidemics, are all likely to influence welfare to a greater extent than will climate change. This is not to say that climate change is unimportant or does not matter. We merely stress that to assess the true significance of climate change it must be evaluated against changes that will occur due to other environmental constraints and social constructs.

Climatologists talking about climate change always run the risk, therefore, of being seen to be slanted in their views. They may be interpreted as being unnecessarily alarmist by those who reckon that human ingenuity and technical change will minimise the effects of climate change (Ausubel, 1995), or overly complacent by those who see climate as a dominant control on human choices and action. Seeing climate, and therefore climate change, as a resource to manage and to benefit from and as essentially neutral is surely a more constructive view to take. The notion of 'good' and 'bad' climates is a hard one to defend in any absolute sense. Temporal changes in climate, seen in this way, present societies with challenges to cope with and opportunities to exploit. These challenges and opportunities presented by climate change are very much those that every colonising community through history have realised are posed by geographical differences in climate. This view of climate, whether implicit or explicit, has been true of, for example, Monguls in Europe, Vikings in Greenland or Europeans in Africa. For example, one may view the nineteenth century history of the interaction between climate and society as one about the ability of the European colonising powers to exploit geographical differences in climate in the Tropics – rubber in Malaysia, cocoa in West Africa or bananas in the Caribbean – and to manage the regional climate impacts that such exploitation might bring with it (Grove, 1994). A twenty-first century history of such interaction may well be about the ability of different communities or regions to exploit and manage the forthcoming temporal changes in climate brought about by human pollution of the atmosphere.

Taking this view, the United Nations Framework Convention on Climate Change is concerned primarily with the regulation of this exploitation and management process as implied in Article 2:

> The ultimate objective of this Convention ... is to achieve ... stabilisation of greenhouse gas concentrations in the atmosphere at a level that would

prevent dangerous anthropogenic interference with the climate system. Such a level should be achieved within a time frame sufficient to allow ecosystems to adapt naturally to climate change, to ensure that food production is not threatened and to enable economic development to proceed in a sustainable manner.

(UNFCCC, 1992)

It is unlikely, however, that such regulation can ensure that communities and nation states benefit or suffer equally from climate change. The partitioning of these benefits or costs between nations will depend on two things: serendipity and access to human and technological capital. The impact of climate change on the world is likely to be dictated largely by the existing inequalities in human vulnerability, with some luck and institutional regulation thrown in.

The view of climate prevailing at the end of the twentieth century is, therefore, as follows. Climate is no longer regarded as a constant, but is continually subject to change. These changes are increasingly being caused, inadvertently, by human behaviour. The changes in climate remain largely unpredictable, but will have important consequences for human welfare, decision-making and planning. In addition to the prospect of climate change, new developments in daily, monthly and seasonal weather forecasting provide an even greater impetus to take climate variability seriously. Organisations, charged with investment decisions, environmental management and future planning strategies, need to include climate in their decision-making structures as a key variable rather than as an assumed constant. If this book contributes to such an awareness with respect to the British Isles, then it will have achieved one of its purposes. It will, in the process, have also contributed to one of the four original aims of the Climatic Research Unit cited when it was established in 1972, namely, 'To investigate the possibilities of making advisory statements about future trends of weather and climate from a season to many years ahead, based on acceptable scientific methods and in a form likely to be useful for long-term planning purposes' (Anon., 1972: 9).

9 Climate of uncertainty

June 1999

An invited commentary in
Environment Action, the newsletter of
the UK's Environment Agency[31]

We know climate is changing. We also know that at least part of the reason is related to human pollution of the global atmosphere. Given that greenhouse gas concentrations in the atmosphere will continue to increase for the forseeable future, we can use climate models to make predictions about future climate. Such predictions form the basis of the growing concern, both nationally and internationally, about the potential dislocations such climate change may cause to our natural and social systems. Nationally, the Department of the Environment, Transport and the Regions has established the UK Climate Impacts Programme (UKCIP) to facilitate assessments of an array of impacts that climate change may have within the UK. This programme of activities will assist in the development of national climate mitigation and adaptation policies. Internationally, the United Nations has established the Intergovernmental Panel on Climate Change (IPCC) to report on the state of knowledge in this area; the Third Assessment Report of the IPCC is now in preparation and is due to be published in early 2001.

But climate prediction is fraught with uncertainty. This uncertainty has two generic sources: unknown, and 'unknowable', future emissions of greenhouse gases and other climate altering aerosols, and incomplete knowledge of the sensitivity of the global climate system to such emissions.

The former uncertainty, unknown future emissions, cannot be resolved through conventional predictive science and is usually accommodated through the construction of a number of scenarios of future world evolution, each of which is associated with a particular greenhouse gas emissions path. The current emissions scenarios under review by the IPCC embody a range of carbon emissions by 2100 from about 6 GtC to over 30 GtC. This range of emissions could lead to global warming values by 2100 of between 1.25° and 4.75°C.

The latter uncertainty, incomplete scientific knowledge of the climate system, is in principle capable of being narrowed, if not fully resolved, through the conventional natural science paradigm (there are two exceptions to this optimistic view which I mention below). For example, more and better observations of cloud dynamics will assist in the improvement of their representation in climate models;

31 Hulme, M. (1999) Climate of uncertainty *Environment Action*, Issue 20 (June/July).

more powerful computing capacity will allow a wider range of processes, and at smaller space-scales, to be simulated by the models; and improved biosphere models, coupled to climate models, will allow us to quantify some of the climate-biosphere feedback processes more accurately.

The two areas where natural science may achieve more limited success in making deterministic predictions of future climate concern the chaotic behaviour of the coupled ocean-atmosphere system and the important feedback between land use change (a socially controlled process) and climate. The climate system is an indeterminate system and, as with numerical weather forecasts over the period 5 to 10 days, the same climate model run with very slightly different initial conditions may yield rather large differences in regional climate predictions. This represents a fundamental limit to our predictive capacity. Similarly, the feedback between land use change and climate brings us back into the realm of social and human behavioural processes, processes that are not deterministically predictable.

For all of the above reasons it is clear that we will never be in a position to predict with certainty future climate, least of all on regional and local scales. Up until now, this situation has been addressed in assessments and reports by adopting one of a number of different approaches: the 'best guess' prediction, a range of predictions, or multiple predictions. IPCC, for example, has tended to favour the 'best guess' approach, coupled to a high and a low estimate to capture (an unspecified part of) the range of possible future climates. In some other climate change impact assessments, such as for UKCIP, multiple scenarios have been presented, four in the case of the national UKCIP climate change scenarios published last October.

I believe, however, that we are moving into a position where we should be able to make more explicit probabilistic statements about future climate change. How likely is it that climate will warm by more than 1.5°C globally? What is the probability that summer precipitation in the UK will decrease? What is the probability that a UK winter will experience fewer than 10 frost days? It is the difference between risk and uncertainty: 'If you don't know for sure what will happen, but you know the odds, that's risk; if you don't even know the odds, that's uncertainty'.[32] Of course, the odds of some of the elements in climate change prediction will remain unknown (e.g. future emissions) and can only be elucidated through expert judgement. Other elements (e.g. regional climate predictability) are, however, capable of yielding more formal probabilities.

This development in the way we talk about future climate change is therefore both possible and necessary. Possible, because of improved techniques for representing uncertainty in climate change scenarios; and necessary, because resource managers and social planners are better able to incorporate such information into decision-making frameworks. Climate scientists and climate change scenario developers are therefore charged with the task of risk assessment;

---

32 Knight, F.H. (1921) *Risk, uncertainty and profit* Hart, Schaffner & Marx, Boston MA.

managers and planners can therefore perform their task of risk management more effectively.

What we know for certain is that the climate future is uncertain. The sooner we quantify this uncertainty, rather than wait for it to be fully resolved, the better we will be able to cope with climate change.

10 There is no longer such a thing as a purely natural weather event

15 March 2000

An essay written for *The Guardian* newspaper[33]

With the immediate crisis in Mozambique gradually receding, now is the time to address the question I have been asked most frequently over these last two weeks, by journalists, friends and acquaintances, and even by my own seven-year-old daughter: 'Were the floods caused by global warming?'

To answer this question, we need to distinguish between the meteorology of what happened – very heavy rains over several days across a very wide area – and the impacts of the rainfall upon river flows and the subsequent human dislocation and suffering. Whether widespread heavy rain leads to extensive flooding depends on local factors such as land cover, dam management and flood control. And whether excessive flooding leads to widespread human suffering depends on where people live, what warning they are given and what options they have to remove themselves from risk. Clearly, there are lessons to be learnt from the events in Mozambique in both of these areas, but they are not lessons directly related to global warming. The more precise question we have to answer is: 'Were the heavy rains over south-eastern Africa caused by global warming?'

This question does not lend itself to a simple 'yes' or 'no' answer. It is not possible to prove or disprove that a given severe weather event only occurred because of a rapidly warming global climate. We can try to argue from first principles by saying that a warmer atmosphere will hold more moisture and, therefore, have the capacity to deliver heavier rainfall. Or we can argue that warmer ocean temperatures potentially provide tropical cyclones with increased energy to make them more vigorous and the rainfall more intense. But in neither case does this convince us that these particular heavy rains were caused by global warming.

Alternatively, we can try to argue from the statistical evidence. Have such intense rainfalls been recorded before in this part of the world? Well, yes they have – in the 1950s, for example. But have such heavy rainfall events become more frequent? For this part of Africa, the data have not yet been analysed with

33 Hulme, M. (2000) There is no longer such a thing as a purely natural weather event *The Guardian*, 15 March, Environment Section http://www.guardian.co.uk/society/2000/mar/15/mozambique.guardiansocietysupplement.

this question in mind and nor is the data particularly long or reliable enough to yield a convincing statistical answer. This analysis, by the way, has been performed in some other regions – the USA, UK and Australia – and evidence has been found for a trend towards more intense daily rainfall totals in recent years. For Mozambique, the definitive answer to our question continues to elude us. Indeed, it is quite possible that, in another 25 years' time, even with the world 0.5°C hotter than today, climate scientists will still be saying the same thing: 'We cannot say that any specific extreme weather event is caused by global warming.'

If we approach our question from a different angle, however, then I believe there is something important to say about the relationship between global warming and extreme weather events – and this needs saying now, and needs repeating each time the need for action to mitigate climate change, or to adapt to its consequences, is questioned. We know that the global climate is warming. We know that the rate of warming has accelerated in recent decades. Following the 1996 report by the UN's Intergovernmental Panel on Climate Change, the balance of evidence reveals a detectable human influence on this warming. A warming global climate will inevitably lead to changes in the behaviour of all weather systems – the systems that actually deliver weather to us, extreme or not. A global climate warmed by human pollution of the atmosphere must yield different magnitudes and frequencies of a whole spectrum of local weather events.

In this sense, therefore, the weather we now experience is the result of a semi-artificial climate; in a fundamental sense, it is different from the weather we would experience on a parallel planet which humans had not polluted. All weather events we experience from now on are to some indeterminable extent tainted by the human hand. There is no longer such a thing as a purely 'natural' weather event.

This is the important message to take away from Mozambique, or indeed from any other extreme weather event, such as the heavy rainfall in Venezuela before Christmas or the windstorms in France after Christmas. We are running massive risks by altering the climate of our planet in ways we do not fully understand, let alone are able to predict with confidence. As with genetically-modified organisms, humanly-modified climates potentially expose us to risks that are largely unknown and unquantified. And the longer we continue to rely on a carbon-based energy economy, the greater these risks will be. It will take more than the 20 per cent reduction in UK greenhouse gas emissions announced last Thursday by the Government to bring these risks under control.

Gambling with our climate may be acceptable to those of us living in affluent societies in the north, reckoning that we can adapt or buy our way out of climate trouble, but this gambling mentality is not one that would find many takers in Mozambique this week.

11 Something to clear the air

18 August 2006

A review of: '*Avoiding dangerous climate change*' edited by John Schellnhuber *et al.*, written for *The Times Higher Education Supplement*[34]

The origins of this book lie in a very specific political goal of the UK Labour Government – to ensure that the G8 Summit at Gleneagles in July 2005 would deliver a progressive international plan of action on climate change. With the USA opting out of the Kyoto Protocol and with continued sniping from the scientific fringe about the realities of climate change, the UK Prime Minister needed a powerful weapon if he was to ensure progress on his much-trailed goal of using his Presidency of the G8 to break the log-jam on international climate change action. It was to science that he turned for such a weapon.

The Exeter Conference on 'Avoiding dangerous climate change' was held at the Hadley Centre from the 1 to 3 February 2005 and brought together about 200 scientists from about 30 countries. It was unusual for a scientific conference to be announced by a Prime Minister, but even more so for him to announce the Conference goals, namely to answer the questions 'what level of greenhouse gases in the atmosphere is self-evidently too much?' and 'what options do we have to avoid such levels?' The primary audience for the Conference deliberations – opinion formers in the USA – is revealed by the Prime Minister's framing of the opening question; only a constitutional lawyer with an eye on American opinion could have suggested that identifying dangerous climate change is as self-evident as the truths that 'all men are created equal' with inalienable rights for life, liberty and the pursuit of happiness.

Does this context for the book suggest that science has been manipulated here to deliver a political goal? Or indeed, that these very unusual circumstances for a scientific Conference compromise the 200 Conference participants or the 108 contributors to the book? I don't believe it does for two related reasons. But what it does do is reveal the dead-end we are fast approaching in the West about converting the scientific narrative of climate change into a globally legitimate and empowering vision for humanity's future on this planet.

---

34 Hulme, M. (2006) *Something to clear the air* Book review of 'Avoiding dangerous climate change' Schellnhuber, H.J. *et al.* (eds) in *Times Higher Education Supplement* 18 August, p.23.

One reason why this volume should not be seen simply as a sophisticated and indirect form of political spinning by Number 10 is that the Conference failed to deliver its political goal. On the one hand, the Conference was regarded by the books editors as a highly successful event 'by any standards and by all accounts'. There is indeed no doubt that for one week in February 2005 climate change was headline news as scientists almost fell over each another in delivering an even more apocalyptic glimpse of the climatic future – the collapse of West Antarctic Ice Sheet (Rapley), the melting of the Greenland ice cap (Lowe *et al.*), the break-down of the Gulf Stream (Schlesinger *et al.*, Wood *et al.*, Challenor *et al.*), the acidification of ocean waters (Turley *et al.*). These were the 'sleeping giants' of the Earth System as reported by Fred Pearce in *New Scientist* that week. And it also explains why *The Independent* newspaper has ever since been running a front-page 'in-your-face' campaign warning all and sundry of the terrors of climate change. On the other hand, did Tony Blair get his Gleneagles break-through by riding the crest of this scientific wave of climate change catastrophes? No he didn't. Although a Gleneagles 'Plan of Action' on climate change was signed by all eight premiers, it did not lead to a sea-change of opinion in Washington and was in fact trumped by George Bush's 'Asia-Pacific Pact on Clean Development', from which the Europeans were excluded, promoting technology-based solutions to climate change.

A second reason why the science reported in this book is not necessarily compromised by its political origins is the very reason why a break-through at Gleneagles was not forthcoming. Contrary to what Tony Blair called for from the Conference there *is* no level of greenhouse gases in the atmosphere which is self-evidently too much and, even if there were, it is not possible for science to pronounce on what that level is. The self-evident truth that all men are created equal and independent is not a truth of science – it is a socially constructed truth that emerges from an enlightened and rational self-conscious reflection on human nature and on humanity's place in the universe. Similarly, what constitutes dangerous climate change will be socially constructed and politically negotiated, through a process in which science will play a role, but in the end a rather limited one. It is by no means science's remit to identify what is self-evidently dangerous.

Now the editors of this book recognise this limitation at an intellectual level – they acknowledge 'it would be expecting too much of the scientific community to act as the arbiter of society's preferences' with regard to dangerous climate change. Yet the content, tone and subliminal message of the book are not fully consistent with this standpoint. For a start, the book only lends voice to selected parts of the scientific endeavour – those who happen to be trained in the natural sciences and in economics. Insights from the social sciences, from ethicists, lawyers and international relations experts about what might constitute dangerous climate change are sadly lacking. Rajendra Pachauri in his Conference keynote speech recognises this when he says 'social scientists have not really got adequately involved in … climate change', so it is a shame the book does not correct this deficiency. The chapter by Yamin *et al.* on operationalising Article 2 of the UN Framework Convention goes the furthest in this direction, but it is a lone voice.

Furthermore, the book only lends voice to a very narrow cohort of the world's scientific community – very largely from the UK and the USA, and almost exclusively from Europe and the English-speaking world. Of the 108 contributors to the book, fully 75 per cent are from the UK and the USA, between them responsible for 60 per cent of the forty-one chapters. If Europe is added, then 85 per cent of the chapters are accounted for. There are three contributors from India, two from Africa, none from China and none from the whole South American continent. It is easy to say of course that this was a Conference called by a British Prime Minister and held in the UK and that the distribution of scientific competences around the world is deeply uneven. This is true. But when the perspective from half of the world's population about 'what constitutes dangerous climate change and how it might be avoided' is represented in an international science-cum-political conference by just three voices one can see the problem.

But the book *Avoiding Dangerous Climate Change* symbolises a still deeper problem which afflicts the struggling project to respond to what the Prime Minister in his Foreword calls 'the world's greatest environmental challenge'. Climate change presents the world with a deep paradox. On the one hand a universal (natural) science has been very effective – as evidenced in part by this volume – at identifying the drivers and risks associated with our transformation of our planetary habitat. The scientific narrative about climate change is strong and growing, and the visions of the future frequently accompanying this narrative are ones that captivate us with a mixture of unease and fear. Yet at the same time, our society – at least Western society – has lost belief in the authority of the meta-narrative. We no longer rally around megaphone calls to live this way or to organise ourselves that way. We are cynical of the powerful, sceptical of the wise, and distrustful of those whom we feel have some one-eyed utopian goal. Our horizons remain limited, our instincts remain parochial, and our institutions poorly designed to cope with global-scale challenges. This is the irony of, for example, the new coalition climate campaign in the UK called *Stop Climate Chaos*. In a socially chaotic – or fragmented – era, where authority structures are weak and contested, we will never be in a position to bring order back to our climate, to stop climate chaos. What we sow in our society we reap in our climate. And science, even science such as presented in this book which suggests plausibly that the stakes are indeed high, reaches its limit to inspire us to re-fashion society in a way that is sustainable.

As Tony Blair has discovered in the 18 months since the Exeter Conference, avoiding dangerous climate change is not a programme to be designed by science. Nor can it be politicised through a party manifesto or delivered by a smart communications campaign. There can be no grand global project in our generation to deliver all our climate change goals – partly because some of these goals are contradictory and partly because the world is now too fragmented to be susceptible to grand projects. We are left – as we have always been left – with our instinct for survival, our ingenuity to adapt, and our peculiar human blend of faith, spirit and sense of destiny. Managing our climate will test all of these virtues.

So if we cannot use the powers of science and economics to force through political action on climate change, are there other alliances or paradigms that can be used to help us 'avoid dangerous climate change'? I believe that there are and a recent book which emerged from a Tyndall Centre conference on dangerous climate change – *Fairness in Adaptation to Climate Change*[35] – begins to point us in the right direction. Seeing climate change as a challenge to our sense of identity and sense of belonging, seeing it as a manifestation of distorted relationships between people and between nations, will in the end be more fruitful than framing it as a techno-economic problem to be solved by some optimising and omnipotent global governor.

What is self-evident about climate change is not that some future level of change is dangerous, but that the human condition, the way we relate to each other and to our natural world, has already put us all at risk.

35 Adger, W.N., Paavola, J., Huq, S. and Mace, M.J. (eds) (2006) *Fairness in adaptation to climate change* MIT Press Cambridge MA.

12 To what climate are we adapting?

30 September 2008

Essay written for the *BBC News On-Line* Green Room[36]

> Moves to adapt our society for a changing climate may have focused rather too much on long-term scenarios and not enough on how to cope with weather and short-term variability, argues Mike Hulme. He says the past two British summers show the dangers of this overemphasis on laudable long-sightedness.

This year's British summer has again failed to meet many of our expectations.

As reported by *The Daily Telegraph* at the end of August: 'Summer is all but over. We are entitled to ask: what summer? For the second year running we have been denied anything worthy of the name.' Rainfall was more than 50 per cent above average; sunshine levels were well below average, with August being particularly gloomy. The average temperature was a fraction above the 1961–90 mean, although this was disguised by cooler days offsetting warmer nights.

Is this the British summer climate to which we are being told we need to adapt?

The Mayor of London's office has just published its first ever climate change adaptation strategy in which the risks of increasing summer heatwaves and droughts compete with those of more frequent winter flooding. Both scenarios demand the attention of Boris Johnson and his urban planners. In a couple of months' time, the government – through the UK Climate Impacts Programme – will launch (with much publicity and scientific acclaim) a new set of climate scenarios called UKCP09 for the nation until the year 2100. They will be the first national scenarios that offer probabilities about how future climate may change.

---

36 Hulme, M. (2008) To what climate are we adapting? *BBC News On-line* Green Room 30 September 2008 http://news.bbc.co.uk/1/hi/sci/tech/7643883.stm.

'Eye off the storm'

As we think more carefully about how we live with our climate and how we can improve our preparedness for future weather, are we over-emphasising long-term prospects over shorter-term realities? Are we paying too much attention to uncertain long-term climate predictions – dominated by greenhouse gas-driven global warming – whilst taking our eye off the more immediate weather futures which will determine the significance of climate for society over the next years and decades? Using the jargon of climate science, are we giving too much weight to the anthropogenic 'signal' of global warming whilst ignoring the natural 'noise' of climate?

Individuals, communities and societies have always wanted to know what the future weather will be; whether for managing the cultivation of land or the building of homes through to preparing for social rituals or communal celebrations. Since the middle of the nineteenth century, scientific weather forecasting has been evolving. Current forecasts are able to predict weather three days ahead. These forecasts are as skilful, and contain more detail, as next-day forecasts 40 years ago. We are better prepared and better adapted to avoid weather risks, such as storms at sea, and to grasp opportunities (for example transport management) than any previous generation. Paralleling these developments, we have in the last 10 to 15 years also been urged to start bringing long-term climate predictions – scenarios for 2050 and 2100 linked to global warming and derived from climate models – into our adaptation planning. This is particularly needed for infrastructure projects which have a long life-span.

Balancing act

But are these long-term climate scenarios alone what we most urgently need to improve society's adaptation to weather and climate – to avoid risks and to grasp opportunities? Weather forecasts offer easily demonstrable and quantified skill. But climate scenarios for the year 2050 cannot be tested against observations; we have to rely on our faith in the underlying climate models. This faith is tested when we endure summers like those of 2007 and 2008.

All long-term climate scenarios suggest British summers will become drier; if we now start adapting for drier summers what happens to farmers, businesses and tourists when we have two successive very wet summers? All long-term scenarios also suggest heatwaves, such as the one in August 2003, will become more frequent, even the norm, by 2050. How does adapting to this prospect improve our ability to survive cool, gloomy weeks like those we had in 2008? We will never know empirically on any useful timescale whether or not we have accurate climate predictions for 2050. Yet even if they do prove accurate, if our shorter-term forewarning of daily weather to decadal climate is poor, we may end up just as maladapted and just as exposed to weather risks as if we had ignored global warming entirely.

Two extremes

Scientists have recently begun to tackle seasonal to decadal climate forecasting, time-scales in which natural variability ('noise') is more important than global warming ('signal'). Yet for now, these forecasts remain primitive and of limited skill. So we remain caught between the two extremes of what science can foretell of future weather: daily forecasts with known skill and value, and centennial scenarios of unknown skill based on (good) faith. We do need to consider the latter in guiding long-term infrastructure design, but an over-reliance on such scenarios to dominate our adaptation thinking and planning carries three dangers.

Long-term climate scenarios may prove to be inaccurate (we have poor means of knowing for sure). Second, even if they do prove accurate on the time-scale of 50 to 100 years, they may be all but useless for foretelling the climate of the next one to 10 years. This is linked to the third danger, which is about the social psychology of weather expectations. Constant public talk of presaged late-century climate will alter public expectations of near-term climate, which – as we have seen these last two years – will continue to yield weather of very different character to that offered by our 2050s scenarios.

To use a specific example: how do we prepare the 2012 London Olympics to be well adapted to British summer climate? Do we take a 2050 climate change scenario – heatwaves, droughts and all – and assume this will best describe the summer of 2012? Do we use one of the new experimental decadal forecasts that suggests we may see little warming and maybe wetter summers over the next decade? Or do we make sure that the Olympics are prepared to cope with whatever the summer of 2012 turns out like – whether the blazing heat of 1995 or the gloom of 2008?

There is an irony here. At the same time as the new national UKCP09 scenarios offer us more detailed information than ever before about the climate of 2050 – for example, probabilities of hourly rainfall at 1 km resolution – so climate science is increasingly emphasising that our weather from months to a decade or two ahead will be dominated by natural variability which we poorly understand and struggle to predict.

The coming decade will yield the familiar mixture of British weather: heat, cold, wind, rain and drought. Yes, let us use such foresight as science can offer us about the longer term; but effective adaptation to weather and climate variability and management of public expectations of future weather demand more than merely this. Premature locking of our infrastructure and our social psychology into the dimly presaged but overly precise climate of the late twenty-first century may be as risky as pretending we are still living with the climate of the twentieth century.

13 Do we need better predictions to adapt to a changing climate?

31 March 2009

Article co-authored with Suraje Dessai, Rob Lempert and Roger Pielke Jr. for *EOS: Transactions of the American Geophysical Union*[37]

Many scientists have called for a substantial new investment in climate modelling to increase the accuracy, precision, and reliability of climate predictions. Such investments are often justified by asserting that failure to improve predictions will prevent society from adapting successfully to changing climate. This Forum article questions these claims, suggests limits to predictability, and argues that society can (and indeed must) make effective adaptation decisions in the absence of accurate and precise climate predictions.

Climate prediction for decision-making

There is no doubt that climate science has proved vital in detecting and attributing past and current changes in the climate system and in projecting potential long-term future changes based on scenarios of greenhouse gas emissions and other forcings. The ability of climate models to reproduce the time evolution of observed global mean temperature has given the models much credibility. Advances in scientific understanding and in computational resources have increased the trustworthiness of model projections of future climates.

Many climate scientists, science funding agencies, and decision makers now argue that further quantification of prediction uncertainties and more accuracy and precision in assessments of future climate change are necessary to develop effective adaptation strategies. For instance, the statement for the May 2008 World Modelling Summit for Climate Prediction[38] argues that, 'climate models will, as in the past, play an important, and perhaps central, role in guiding the trillion dollar decisions that the peoples, governments and

---

37 Dessai, S., Hulme, M., Lempert, R. and Pielke, R. Jr. (2009) Do we need better predictions to adapt to a changing climate? *EOS* 90(13), 111–112.
38 See: http://wcrp.ipsl.jussieu.fr/Workshops/ModellingSummit/Documents/FinalSummit Stat_6_6.pdf (accessed 22 December 2012).

industries of the world will be making to cope with the consequences of changing climate.'

The statement calls for a revolution in climate prediction because society needs it and because it is possible. The summit statement argues that such a revolution 'is necessary because adaptation strategies require more accurate and reliable predictions of regional weather and climate extreme events than are possible with the current generation of climate models.' It states that such a revolution is possible because of advances in scientific understanding and computational power.

If true, such claims place a high premium on accurate and precise climate predictions at a range of geographical and temporal scales as a key element of decision making related to climate adaptation. Under this line of reasoning, such predictions become indispensable to, and indeed are a prerequisite for, effective adaptation decision making. Until such investments come to fruition, according to this line of reasoning, effective adaptation will be hampered by the uncertainties and imprecision that characterise current climate predictions.

Limits of climate prediction

Yet the accuracy of climate predictions is limited by fundamental, irreducible uncertainties. For climate prediction, uncertainties can arise from limitations in knowledge (e.g., cloud physics), from randomness (e.g., due to the chaotic nature of the climate system), and from human actions (e.g., future greenhouse gas emissions). Some of these uncertainties can be quantified, but many simply cannot, leaving some level of irreducible ignorance in our understanding of future climate.

An explosion of uncertainty arises when a climate change impact assessment aims to inform national and local adaptation decisions, because uncertainties accumulate from the various levels of the assessment. Climate impact assessments undertaken for the purposes of adaptation decisions (sometimes called end-to-end analyses) propagate these uncertainties and generate large uncertainty ranges in climate impacts. These studies also find that the impacts are highly conditional on assumptions made in the assessment, for example, with respect to weightings of global climate models (GCMs) – according to some criteria, such as performance against past observations – or to the combination of GCMs used.

Future prospects for reducing these large uncertainties remain limited for several reasons. Computational restrictions have thus far restricted the uncertainty space explored in model simulations, so uncertainty in climate predictions may well increase even as computational power increases. The search for objective constraints with which to reduce the uncertainty in regional predictions has proven elusive. The problem of equifinality (sometimes also called the problem of 'model identifiability') – that different model structures and different parameter sets of a model can produce similar observed behaviour of the system under study – has rarely been addressed. Furthermore, current projections suggest that the

Earth's climate may soon enter a regime dissimilar to any seen for millions of years and one for which paleoclimate evidence is sparse. Model projections of future climate therefore represent extrapolations into states of the Earth system that have never before been experienced by humanity, making it impossible to either calibrate the model for the forecast regime of interest or confirm the usefulness of the forecasting process.

In addition, climate is only one of many important processes that will influence the success of any future adaptation efforts, and often it is not the most important factor. Our current ability to predict many of these other processes – such as the future course of globalisation, economic priorities, regulation, technology, demographics, cultural preferences, and so forth – remains much more limited than our ability to predict future climate. This raises the question of why improved climate predictions ought to be given such a high priority in designing adaptation policies.

Alternatives to prediction

Individuals and organisations commonly take actions without having accurate predictions of the future to support those actions. In the absence of accurate predictions, they manage the uncertainty by making decisions or establishing robust decision processes that produce satisfactory results. In recent years, a number of researchers have begun to use climate models to provide information that can help evaluate alternative responses to climate change, without necessarily relying on accurate predictions as a key step in the assessment process. The basic concept rests on an exploratory modelling approach whereby analysts use multiple runs of one or more simulation models to systematically explore the implications of a wide range of assumptions and to make policy arguments whose likelihood of achieving desired ends is only weakly affected by the irreducible uncertainties.

As one key step in the assessment process, such analyses use climate models to identify potential vulnerabilities of proposed adaptation strategies. These analyses do not require accurate predictions of future climate change from cutting-edge models. Rather, they require only a range of plausible representations of future climate that can be used to help organisations, such as water resources agencies, better understand where their climate change-related vulnerabilities may lie and how those vulnerabilities can be addressed. Even without accurate probability distributions over the range of future climate impacts, such information can prove very useful to decision makers.

Such analyses generally fall under the heading of 'robust decision making.' Robust strategies perform well compared with alternative strategies over a wide range of assumptions about the future. In this sense, robust strategies are insensitive to the resolution of the uncertainties. A variety of analytic approaches, such as exploratory modelling, have been proposed to identify and assess robust strategies.

Climate and science policy implications

Given the deep uncertainties involved in the prediction of future climate, and even more so of future climate impacts, and given that climate is usually only one factor driving the success of adaptation decisions, we believe that the 'predict-then-act' approach to science in support of climate change adaptation is significantly flawed. This does not imply that continued climate model development cannot provide useful information for adaptation. For instance, such development could further inform the plausible range of impacts considered when crafting a robust adaptation strategy. However, further scientific effort will never eliminate uncertainty; it may in fact increase uncertainty. For example, three decades of research on climate sensitivity (the global mean temperature change following an instantaneous doubling of carbon dioxide in the atmosphere) have not reduced, but rather have increased, the uncertainty surrounding the numerical range of this concept. The lack of climate predictability should not be interpreted as a limit to preparing strategies for adaptation.

By avoiding an analysis approach that places climate prediction at its heart, successful adaptation strategies can be developed in the face of deep uncertainty. Decision makers should systematically examine the performance of their adaptation strategies over a wide range of plausible futures driven by uncertainty about the future state of climate and many other economic, political, and cultural factors. They should choose a strategy they find sufficiently robust across these alternative futures. Such an approach can identify successful adaptation strategies without accurate and precise predictions of future climate.

Our arguments have significant implications for science policy. At a time when government expects decisions to be based on the best possible science (e.g., evidence-based policy making), we suggest that climate science is unlikely to support prediction-based decisions. Over-precise climate predictions can also lead to mal-adaptation if the predictions are misinterpreted or used incorrectly. From a science policy perspective, it is worth reflecting on where investments by science funding agencies can best increase the societal benefit of science. Efforts to justify renewed investments in climate models based on promises of guiding decisions are misplaced.

The World Modelling Summit for Climate Prediction called for a substantial increase in computing power (an increase by a factor of 1000, at the cost of more than a billion dollars) to provide better information at the local level. We believe, however, that society will benefit more from having a greater understanding of the vulnerability of climate-influenced decisions in the face of large irreducible uncertainties, and the various means of reducing such vulnerabilities, than from any plausible and foreseeable increase in the accuracy and precision of climate predictions.

14 On the origin of the greenhouse effect: John Tyndall's 1859 interrogation of Nature

May 2009

Article published in *Weather*, a magazine of the Royal Meteorological Society, to mark the 150th anniversary of John Tyndall's experiments demonstrating the absorptive properties of greenhouse gases[39]

One hundred and fifty years ago this month, on Wednesday 18 May 1859, after a full day's work in the basement laboratory at the Royal Institution in central London, the 38 year-old Irish scientist John Tyndall wrote in his journal: 'Experimented all day; the subject is completely in my hands!' Tyndall had been experimenting with the absorptive properties of gases and vapours with a view to testing the idea that different gases, commonly found in the Earth's atmosphere, absorbed differing amounts of radiant heat (i.e., long-wave infra-red radiation). This was the idea that the physicist Joseph Fourier had articulated more than thirty years earlier and which Claude Pouillet had more recently elaborated. But no-one had demonstrated experimentally such a phenomenon existed. If true, the idea could help explain how the temperature of this, and other planets, was regulated. Tyndall's experiments in May 1859, which he further refined for several years afterwards, established for the first time that molecules of gases such as water vapour, carbon dioxide and methane, do indeed absorb more energy than oxygen and nitrogen when radiant heat is passed through them.

Less than four weeks after declaring success in his journal, on Friday 10 June at an evening meeting of the Royal Institution with Prince Albert, the Prince Consort, in the chair, John Tyndall offered the first public experimentally based account of what has become known as 'the greenhouse effect'. This physical basis for anthropogenic global warming was established six months before Charles

---

[39] Hulme, M. (2009) On the origin of the greenhouse effect: John Tyndall's 1859 interrogation of Nature *Weather* 64(5), 121–123.

Darwin published *On the Origin of Species* which laid the biological basis for life's evolutionary diversity. This short article offers an account of Tyndall's ground-breaking experiments 150 years ago and remarks on their significance for later discussions about climate change.

'On account of certain speculations ...'

John Tyndall was one of the outstanding scientific personalities of the Victorian age and a passionate defender, promoter and populariser of the scientific naturalism which flowered in later Victorian Britain (Turner, 1981). In a testimonial essay in June 1887 to mark his retirement, the journal *Nature* remarked that,

> Others will rank beside or above [Tyndall] as investigators, but in the promotion of the great scientific movement of the last fifty years he has played a part second to none ... uniting scientific eminence of no ordinary kind with extraordinary gifts of exposition, he has, by his lectures and his books, brought the democracy into touch with scientific research.
>
> (Anon., 1887: 217)

In the era before professional scientific training, Tyndall's education and early career had been typically eclectic. After his schooling in County Carlow, Ireland, he became a junior railway surveyor in the north of England during the railway boom years of the 1840s, before teaching school mathematics in Hampshire and then earning his doctorate in experimental chemistry at Marburg in Germany in 1850. His professional career in science was finally established back in Britain when Michael Faraday invited him, in 1853, to give a series of discourses at the Royal Institution in London. The following year he was there appointed Professor of Natural Philosophy. Under Faraday's patronage, Tyndall was to develop and display his talent for experimental science applied across an astonishing variety of subjects (see Eve and Creasey, 1945, Friday *et al.*, 1974 and Brock *et al.*, 1981 for the best accounts of his life and career).

Tyndall's interest in radiant heat and its passage through the atmosphere was triggered by his long-standing interest in glaciers and their mass balance. He was a keen Alpine mountaineer and in 1858 became a member of the newly founded Alpine Club. This curiosity about glaciers had brought him into contact with the earlier ideas of De Saussure, Fourier, Pouillet and Hopkins regarding the differential passage of solar and terrestrial radiation through the atmosphere. The French physicist Claude Pouillet, for example, had published a memoir in 1838 – 'Memoir on solar heat, on the radiating and absorbing powers of atmospheric air, and on the temperature of space' (Pouillet, 1838) – in which he had attached the properties of water vapour and carbon dioxide in the atmosphere to Fourier's

1820s theory of terrestrial temperatures.[40] It was 'on account of certain speculations' of these scientists that Tyndall started his experimental work on radiant heat. As Tyndall explained: 'It was supposed that the rays from the sun and fixed stars could reach the earth through the atmosphere more easily than the rays emanating from the earth could get back to space' (Tyndall, 1859a: 156). For Tyndall 'this view required experimental verification' and as an experienced and creative experimenter he set to work in the laboratory of the Royal Institution in the spring of 1859.

'Putting his questions to Nature ...'

John Tyndall was already a familiar and well-known scientist among the growing number of scientific clubs, societies and institutions of London in mid-Victorian Britain (Lightman, 2007). He was made a Fellow of the Royal Society in 1852 and, from 1856, a member of its Council. He regularly gave public lectures on science at the London Institution and at the Royal Institution; he was a member of the Philosophical Club and the Alpine Club and, in 1859, was elected to one of London's leading clubs for public intellectuals, the Atheneum. He also held a professorship of physics at the Royal School of Mines.

Amidst this busy public and professional life Tyndall set himself to work on devising the experimental apparatus that would allow him to test the ideas of Fourier and Pouillet. By early May of 1859 he had made good progress with his experimental design and was able to write in his journal[41] of Monday 9 May: 'Tested apparatus for the transmission of radiant heat through various gases. Tried various experiments, but as yet no effect has been observed.' He clearly sensed, however, that the 'questions he was putting to Nature' (Tyndall, 1859a: 157) were close to being answered and he was working long hours – between eight and ten hours a day (Tyndall, 1861) – in the basement laboratory of the Royal Institution on Albemarle Street, about 40 minutes' walk from his home in Maida Vale. On Wednesday 11 May he was busy enough to eat his 'tea in the laboratory' and he spent much of the following Sunday 'devising experiments'. And he was experimenting again on Tuesday 17 May, the evening of which he spent relaxing at the home of his Alpine climbing companion, the chemist Edward Frankland.

The following day – Wednesday 18 May 1859, the culmination of seven weeks of intense experimenting – was the signature day for his project as he recorded in his journal: 'Experimented all day; the subject is completely in my hands!' Tyndall was in high spirits following this success and that evening he dined at the house of Colonel Yorke, returning home later that evening 'arm in arm' with Frankland

---

40 See pp.55–64 in Fleming, J.R. (1998) for a good discussion about the early provenance of claims about 'the greenhouse effect'.

41 In this article all quotations from Tyndall's journal are extracted from the relevant pages in Tyndall (1855–1872).

and another of his friends, Richard Dawes the Dean of Hereford.[42] Tyndall recorded in his journal a trace of the jovial spirit of the evening: 'On passing the Archbishop of York's [house] Dawes said he was to dine there on the following day – I said he would be out of this element, which he admitted'. Yet he was back at his work the following day, 'experimenting, chiefly with vapours, coal gas wonderful – ether vapour still more so'.

Tyndall was fully aware at this time of the potential significance of his experimental findings, findings which lent support to the theoretical speculations of earlier physicists. Just a few days later – on 1 June – he wrote to one of his correspondents, the German physicist and mathematician Rudolf Clausius, outlining the purpose and results of these experiments (Tyndall, 1859b). And just over three weeks later – on the evening of Friday 10 June – he demonstrated and delivered a lecture at the Royal Institution which was to be the first public exposition of his findings and in which he also outlined their significance. Prince Albert the Prince Consort – Queen Victoria's husband, Vice-Patron of the Royal Institution and great supporter of science and engineering – was in the chair for this lecture and he would have seen and heard Tyndall explaining the results of his interrogation of nature. 'The bearing of this experiment upon the action of planetary atmospheres is obvious ... the atmosphere admits of the entrance of the solar heat, but checks its exit; and the result is a tendency to accumulate heat at the surface of the planet' (Tyndall, 1859a: 158). Fourier's and Pouillet's regulation of planetary temperatures through the differential actions of atmospheric gases now had experimental evidence in its favour. Tyndall's remarks also evidence his high view of laboratory experimentation: 'But whatever be the fate of speculation, the experimental fact abides – that gases absorb radiant heat of different qualities in different degrees' (Tyndall, 1859a: 158).

With this initial success behind him, Tyndall now for a while became distracted from follow-up work. He was working hard to complete his first book – *The Glaciers of the Alps* – to be published the following June (Tyndall, 1860) and he also entered into a round of correspondence and conversations with many friends and colleagues about an approach from the University of Edinburgh head-hunting him for the recently vacated position of Professor of Natural Philosophy. This was an offer he eventually turned down, contrary to the advice of many of his friends and despite the offer of a salary of £1,250, four times what he was earning at the Royal Institution (Forgan, 1981) and the equivalent to an annual earning potential today of about £85,000.

By September 1860, however, with these distractions behind him, Tyndall returned to his radiant heat experiments. He wanted to improve upon the accuracy of his measurements and to extend the range of gases and vapours tested beyond the ambient air, water vapour, hydrogen, coal gas (a mixture of hydrogen, methane, ethylene and carbon monoxide), bisulphide of carbon (carbon

---

42 Despite Tyndall's widely known atheistic beliefs, he maintained many friendships with members of the church establishment.

disulphide) and carbon dioxide he had tried in 1859. His biggest problem was to secure a reliable and constant heat source. The heated copper plates he had used in 1859 were unsatisfactory and now, after trying with a cube of hot oil, he eventually settled on the solution of a Leslie Cube – a metal cube containing boiling water the temperature of which Tyndall found he was able to control to his satisfaction.

'All the mutations of climate ...'

Excited about these new results, in January 1861 Tyndall prepared a paper on the action of gases and vapours on radiant heat for submission to the Royal Society for their prestigious annual Bakerian Lecture (Tyndall had already delivered this lecture in 1855 – on magnetic forces – and he was to be so honoured again in 1864 and 1881). His paper on radiant heat was selected by the Society – a sign of his eminent standing in Victorian scientific circles – and on Thursday 7 February 1861 he delivered his lecture in the premises of the Royal Society in Burlington House, Piccadilly, just two minutes' walk from his laboratory in the Royal Institution. Tyndall did so feeling seriously ill. That morning he had risen, 'with a head slightly heavy and against my general convictions I ... took about half a teaspoonful of Bell's citrate of magnesia. For the last six months [I] did not feel as much out of order as [I did] during the entire day. Damn physic!' (Tyndall journal, 7 February 1861). The lecture 'On the absorption and radiation of heat by gases and vapours, and on the physical connexion of radiation, absorption and conduction' – later published as Tyndall (1861) – was a great success: 'I never saw so large an attendance in the rooms of the Society – I spoke for about an hour and a half, and they heard me with breathless attention from beginning to end. There were many remarkable men present. [Lord Alfred] Tennyson [the Poet Laureate] was there; saw him afterwards – he wanted me to smoke a pipe with him, but I feared my brain' (Tyndall journal, 7 February 1861).

John Tyndall's experimental work in 1859, and again 1860–1861, became central for subsequent understanding of the heat budget of the atmosphere. Tyndall demonstrated that a group of polyatomic gases – a group later collectively named 'greenhouse gases' – possessed distinct and differential radiative properties with regard to infra-red radiation. 'These experiments furnish us with purer cases of molecular action than have been hitherto attained in experiments of this nature' (Tyndall, 1861: 277). His work suggested the possibility that by altering concentrations of these gases in the atmosphere human activities could alter the temperature regulation of the planet. Tyndall could forsee such a possibility already in 1861. Changes in the amount of any of the radiatively active constituents of the atmosphere 'may have produced all the mutations of climate which the researches of geologists reveal' (Tyndall, 1861: 277), although he was a long way from developing a coherent account of how human actions could induce significant changes in climate on much shorter time-scales.

'The enormous importance of this question ...'

Tyndall's research was demonstrated and published at the beginning of the intellectually tumultuous decade of the 1860s. On Thursday 24 November 1859, just six months after Tyndall's initial experimental results, Charles Darwin's book *On the Origin of Species* was published in London. Along with this challenge to the prevailing orthodoxy of a fixed biological creation, scientists were also still grappling with the equally revolutionary implications of Louis Agassiz's 1837 'ice age theory'. Trying to understand the causes of these great ice age fluctuations in climate was one of the issues of the time. The ideas of Darwin and Agassiz were assaulting fundamental conceptions of time and stability in, respectively, biological and climatic history.

John Tyndall was intimately connected with these debates. He became a close friend of Thomas Huxley, Joseph Hooker and other scientists in Darwin's circle and was one of the members of the X-Club, an exclusive scientific dining club of nine members founded in 1864 and out of whose discussions the journal *Nature* was published in 1869. He was also consulted by Charles Lyell – who was trying to evaluate Croll's newly published orbital theory – about whether Tyndall's new radiative theory of climatic change could help unravel the causal mystery of Agassiz's ice ages. On 1 June 1866, in a reply to an approach from Lyell, Tyndall stated that changes in radiative properties alone were unlikely to be the root causes of glacial epochs (Fleming, 1998: 74), thus somewhat contradicting his earlier supposition of 1861 quoted above. These exchanges between Lyell and Tyndall about theories of climatic change presaged much later arguments about the interplay between natural and human factors in the modification of global climate.

John Tyndall is deservedly credited with establishing the experimental basis for the putative 'greenhouse effect', first suggested by De Saussure in the 1770s and then developed by Fourier in the 1820s. Svante Arrhenius, for example, in his classic 1896 paper which introduced the idea of the climate sensitivity of carbon dioxide in the atmosphere, praised Tyndall for 'point[ing] out the enormous importance of this question' (Arrhenius, 1896: 237). Gilbert Plass in his 1956 article in *Tellus* which gave widespread visibility and impetus in the post-War era to the carbon dioxide theory of climate change also quoted from Tyndall's 1861 Bakerian Lecture at Burlington House and recognised him as the first 'to attempt a calculation of the infra-red flux of the atmosphere' (Plass, 1956: 140). Tyndall was also correct in identifying the fundamental role of water vapour in atmospheric dynamics which, he claimed, 'must form one of the chief foundation-stones of the science of meteorology' (Tyndall, 1863: 54).

The differential radiative absorption properties of the gases and vapours revealed by Tyndall's interrogation of nature 150 years ago this month – a suite of gases now expanded to include a group of artificial gases unknown to Tyndall, the halocarbons – remain central to the idea of anthropogenic climate change. Subsequent work has established the global warming potentials of each of these gases with some level of precision (e.g. Foster *et al.*, 2007), calculations that are pivotal in efforts to quantify the extent of human influence on the world's temperature and in efforts to reduce and manage those effects. John Tyndall's experimental work in 1859 in the basement of a renowned London scientific institution may not be remembered in the same way as is Darwin's masterpiece *On the Origin of Species* and yet in its own way the legacy of Tyndall's work is just as significant for contemporary cultural and scientific debates.

Section three

Researching

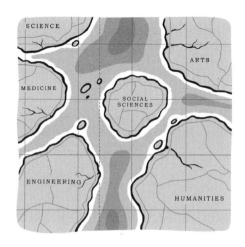

Introduction

This section contains six items which in different ways reflect on the institutions and practices of disciplinary and inter-disciplinary climate change research. The first two items are unpublished and both relate to the Tyndall Centre for Climate Change Research which I established in 2000 and led for its first seven years. Chapter 15 is the speech I delivered at the official opening of the Tyndall Centre in November 2000 to an audience of 300 scientists, representatives of organisations and businesses, research funders and government officials. It offers a view of what I then thought the main aims and purpose of the Centre would be. The next item (Chapter 16) comes from five years later – May 2005 – as the Centre was engaged in an extended process of re-funding by the UK research councils. This chapter is the transcript of an interview with an MPhil student who was researching the institutional settings of climate change research. The transcript is interesting for revealing the perspectives on climate change research I had gained while running the Tyndall Centre, but also some of the frustrations encountered in the struggle to secure continued research funding. In particular I allude to the different and conflicting expectations of UK research councils, universities, government and other social institutions which had to be negotiated and reconciled.

By the time the next item was published in March 2007, I was close to ending my time as Tyndall Centre Director and I was increasingly critical of the conventional communication (see Chapter 37) and policy framing (see Chapter 30) of climate change. This article (Chapter 17) started off as an invited book review for the environment section of *The Guardian* newspaper, but turned into an account of 'post-normal science' and its relevance for understanding controversies around climate change. This article has been widely cited as evidence that I am offering 'a peculiar and incoherent argument about science being based on values and ideology'.[43] I leave it for readers to judge the validity of such a claim.

43 This quote is taken from Clive Hamilton's post 'Climate change and the soothing message of luke-warmism' for *The Conversation* web-site (25 July 2012) http://theconversation.edu.au/climate-change-and-the-soothing-message-of-luke-warmism-8445 (accessed 19 December 2012).

After leaving the Tyndall Centre in 2007, I set out very deliberately to widen the disciplinary perspectives being brought to the study of climate change, both in my own work but also through stimulating the work of others in the academy. My 'boundary crossings' essay published in the leading geography journal *Transactions of the Institute of British Geographers* at the end of 2007 (Chapter 18) contributed to this ambition. It sought to encourage geographers to engage with the idea of climate change from a deeply cultural perspective, complementing if not challenging dominant scientific accounts of the phenomenon. Two years later, early in 2010, I was in a position to launch a new academic review journal – *Wiley Interdisciplinary Reviews (WIREs) Climate Change* – to further advance this ambition. Chapter 19 is a lightly edited version of my editorial essay which launched the journal. It offers my reasons for bringing together in one web-based journal publication authoritative reviews of climate change written from disciplines as diverse as philosophy, oceanography, literary criticism, climatology, social psychology and ecology.

My belief in the need to engage the full spectrum of disciplines in the study and interpretation of climate change – what climate change *means* for human beings – is reflected in the sixth and final item in this section (Chapter 20). This is an essay written in 2011 for another newly launched journal – *Nature Climate Change* – in which I set out a case for the importance of the humanities and interpretative social sciences for understanding climate change. I argue for elevating the role of story-telling alongside that of fact-finding and for appreciating that the challenges of climate change will be resolved as much through cultural change as solved by science and technology.

15 Launching the Tyndall Centre

9 November 2000

Unpublished speech at the opening of
the UK's Tyndall Centre for Climate
Change Research[44]

Iconic and rhetorical it maybe, but the essence of this future vision we have just shown is *choice* – choice about lifestyle, choice about energy systems, choice about transportation strategy. Our studies of climate change have taught us over these last two decades not only that we are already altering global and regional climates through our energy, industrial and land use regimes, but that there are choices to be made about how these regimes develop ... and that these choices will shape the climate future for the next one, two, or more generations.

But there are always good choices or bad choices to be made. To make good choices we need two things: a) a wide range of options from which to choose, and b) a clear understanding of the consequences of each option. It is here that I believe the Tyndall Centre, now officially opened, will make its main contribution to facing the challenge of climate change. Over the longer term we should be judged on whether the Tyndall Centre has helped develop a more sustainable relationship between climate and society through offering a wider range of response options and through evaluating the consequences of each option for this and future generations.

The Tyndall Centre is first and foremost concerned with research. Through our nine partner institutions – University of East Anglia, University of Manchester Institute of Science and Technology, University of Southampton, the universities of Cambridge, Sussex, Cranfield and Leeds, Rutherford Appleton Laboratory and NERC's Centre for Ecology and Hydrology – we harness a wide range of UK expertise across the environmental, economic and engineering disciplines and deploy this expertise to tackle some of the barriers to understanding and action related to climate change. I will say more about Tyndall Centre research in a minute.

But the Tyndall Centre is *more* than just a research centre. We are committed – as our Vision Statement says – to help train the next generation of young

44 The Tyndall Centre was formally opened by Michael Meacher, Minister of State for the Environment in the UK Government, at a ceremony held at the University of East Anglia, Norwich, on 9 November 2000. This speech is lightly edited to suit a printed format rather than a verbal delivery.

scientists and international decision-makers; to enter into public dialogue about what choices society can make to safely manage climate change; and to work with stakeholder and business organisations to advance practical and sustainable responses to climate change. That is why, for example, we are showcasing today the new petrol/electric hybrid car from Toyota just launched on the UK market and which more than halves carbon dioxide emissions compared to conventional vehicles. And in setting these objectives we follow the lead of John Tyndall, the nineteenth-century Victorian inventor, scientist and public educator after whom we have named the Centre, and who was the first to prove the energy absorbing properties of carbon dioxide and other greenhouse gases.

In conducting these various outreach activities, the Tyndall Centre will work alongside and in partnership with many of the organisations in both public and private sectors that you represent here today. We look forward in the months to come to strengthening our existing collaborations and to developing new ones. Furthermore, this agenda is not just for the UK, but worldwide – through collaboration with some of the leading science and policy initiatives in Europe, in North America and in the developing world. The Tyndall Centre, for example, will play a key role in the newly emerging European Climate Forum – a partnership between senior science, business and government bodies in Europe that will facilitate the open exchange of ideas and the pooling of knowledge on how best to respond to climate change.

Let me say a few further words about the big research questions we face and how the Tyndall Centre will advance our understanding in these areas. I believe that there are three key questions that lie at the heart of how we should respond to our changing climate.

First: how can we determine what rate of climate change poses appreciable dangers to human health and to the environment? Danger implies unacceptable risk. And to assess risk we need to know both the probabilities of particular outcomes and the consequences of those outcomes. The reality is that at the moment we don't know what these risks are. For example, is the collapse of the West Antarctic Ice Sheet (a low-probability, high-impact event) a greater risk to society than a 1°C global warming (a high-probability, low-impact event)? The Tyndall Centre can help quantify these risks and assist society evaluate what level of risk it is willing to accept. This will require input from social scientists, as well as from climate scientists, engineers, and biologists engaging with society at large.

The second question is: can we control climate sufficiently to prevent warming reaching this 'dangerous' level? It is the question that lies at the heart of the debate regarding the emissions targets set at Kyoto. But can the Kyoto target, or indeed any other target such as a 60 per cent reduction proposed by the UK's Royal Commission on Environmental Pollution, be reached? Although we know that people will have to change the way they live and businesses change the way they make money, we don't know how easy this will be to achieve. We need to be able to answer such questions as why are some Western societies so reluctant to consider nuclear power as an alternative to fossil fuels? Or why are we so opposed to taxes on fuel, even when we know that some of the money will be spent on energy efficiency improvements? We need more evidence-based policy and, again, the Tyndall Centre will make an important contribution to the debate.

Climate research is also critical for the third question: what happens if we cannot reduce the rate of climate change enough to avoid unacceptable risk? Can we adapt our institutions, our regulations, our behaviour to accommodate the unavoidable dimensions of climate change? I believe identifying 'insurance policies' that we may need to cope with climate change is an entirely rational response to the problem. Here again, the broad range of scientific expertise offered by the Tyndall Centre can help put the necessary debate on a sound footing. As the current floods in this country have shown, adapting to climate change is not so much an issue of technical ability, but rather a question of identifying socially and economically acceptable precautionary policies, and implementing them.

My responses to these three questions shows that to shape a global or a UK society that can evolve with a changing climate, we need to strengthen links between knowledge producers and policy processes, between research and action. I argue that the real challenge for climate change science in the end is not simply to predict future climate; rather, it is to give society the options to choose its own climate future. And to choose well.

And finally, let me announce that to complete our team in the Tyndall Centre, we hope to appoint in the very near future Professor John Schellnhuber, currently Director of the Potsdam Institute for Climate Impact Research, to be the Tyndall Centre's first Research Director. Professor Schellnhuber unfortunately is unable to attend the launch today owing to a long-standing commitment to an important meeting in Germany, but he sends a message of support to the Tyndall Centre today. Let me quote from this message directly:

> It is exciting that the UK is establishing this new research centre which among many things will make a major contribution to the next generation of integrated assessment models for climate change. This contribution will be further enhanced through integrating, from the outset, stakeholder wisdom and perspectives in the research strategy. The Tyndall Centre extends the concept of a focused single-site research institute into a more inclusive national network of research expertise. I have every confidence that the Tyndall Centre will become a world-leading inter-disciplinary research centre for climate change. I wish you well today for the launch of the Centre.

Thank you again for coming today and for supporting us as we launch this important new research enterprise. Thank you especially to Sir Geoffrey Allen for hosting this event, to Professor Lawton and Sir Anthony Cleaver for their words of support, and most importantly to the Minister for the Environment, the Right Hon. Michael Meacher, for opening the Centre today. Thank you.

16 The Tyndall Centre, inter-disciplinary research and funding

May 2005

Unpublished research interview for a
University of Cambridge M.Phil. thesis[45]

My career and the Tyndall Centre

Q The first question that I would ask you concerns your educational background and how exactly you got into climate change research?

A My first degree was in geography specialising in climatology. I then moved on to complete a three-year Ph.D., again in a geography department, on the topic of rainfall variability in Sudan. I graduated with my Ph.D. in 1985 and took up a lecturing position in physical geography at the University of Salford where I undertook a range of teaching and research duties. I moved to the University of East Anglia [UEA] in 1988 where I took up a contract researcher position in the Climatic Research Unit [CRU). So I went from a permanent lectureship to a temporary research contract, which is the opposite of what most people do. And the specific remit that I had in the CRU job was to work on evaluating the performance of global climate models – how well global climate models replicated observed climates [see Chapter 5, this volume].

Q Did that have any sort of influence on the idea for the Tyndall Centre?

A Not directly. The idea for the Tyndall Centre originated within the UK research councils who secured £10 million from the UK Treasury in 1998 to invest in a new initiative that would look at various aspects of climate change. The research councils established a competition amongst universities to come up with the best operation, the best vision of how that would work. So I got drawn into this because UEA with its reputation for climate change research was in a good position to lead a consortium bid. And I got drawn into this, in

---

45 This interview was conducted and recorded in my office in the Tyndall Centre at the University of East Anglia and was part of the data collection for the following thesis: Stiska, A. (2005) *Managing an uncertain debate: global climate change and its research collaborations* Unpublished M.Phil. Thesis in Modern Society and Global Transformations, Faculty of Social and Political Science, University of Cambridge. The interview has been lightly edited to make it readable in printed form and section headings have been added.

a sense, because … why did I get drawn into it? I suppose it was because I had a good grasp of a range of issues, from both physical climate science to social science. Geography is very good training for that type of versatile, flexible thinking. My first connection to climate policy had been through a series of contracts that I had undertaken for Defra.[46] So, for various reasons I found myself being pulled into the new Centre.

Climate model evaluation and scenarios

Q You've obviously kept up with geography research. Or, has that occurred only since you entered the Tyndall Centre?

A I still sometimes identify myself clearly as a geographer, although I haven't worked within the geography discipline formally for many years. But I'm still proud of my geography background. On the research side … I moved from evaluating climate models into, well that led me in a number of directions. One of them is that in order to evaluate climate models, you have to have good observational data, benchmark data against which you're testing. So, quite a lot of my research then moved into how we get good quality data about the real-world climate to test the models against.

And then that also took me into this whole issue of, well, if we're going to use results from climate models to reveal future climate possibilities, then we're into the business of creating scenarios for the future. And so quite a lot of my work then moved into the whole question of designing climate scenarios, to assess the impact of climate change, to inform policy options. So, really my reputation and my invitation to join the IPCC in 1996 was because of my specialist expertise on climate scenarios. Climate scenarios fit in this interface between climate models and real-world data, but they also fit between science and policy, because scenarios are used to illuminate policy choices or to elaborate the consequences of different policy options.

Q The models that you've made, if you're trying to establish the best kind of data that goes into them, are still incomplete you've said. Does the process of matching them to real world data make them more believable?

A I'm not a model developer. The models that we were testing or evaluating were the models that were developed in particular by the UK Met Office, but also the other leading models from the United States, Europe, Japan. We didn't do any model development. But what we were doing, and one of the reasons that Defra gave us a series of contracts over a period of time, was to have some 'independent' evaluation of the Met Office Hadley Centre model. And part of the reasoning was that models needing testing against real-world data to identify those things at which that they were skilful. My objective was to find

---

46 The UK's Department for Environment, Food and Rural Affairs.

quantitative scores, measures of performance. I focused particularly in my work on precipitation.

The question then was how do you feed that information back. If you say 'this model is crap', how do the model developers take that on board? So you then have to develop a relationship with the model developer. And some of our results clearly were just that. We reinforced the notion that some aspect of the model was good, but there were other reasons why aspects of the model weren't so good. It also, of course, raises the question of how good does a model have to be before you can use it for policy planning purposes. Because I think with climate models, to have a perfect representation of the climate system, dream on, you're never going to get that. The question is how good does the model have to be before it starts getting used to inform policy. That, of course, is one of the questions that is really at the centre of the debate about climate change: how good do those models have to be? You can pull out any number of answers to that question and people do!

Q But that's mainly been handled by asking members of the Government or Parliament what level of performance – if it's 80 per cent perfected, 90 per cent perfected, or whatever number you assign to it – would be acceptable for policy-making?
A Well, I don't think it's as clear at that. There are various processes and assessments that are put into place to make that connection between science and policy. Clearly, policy advisors or politicians are not well equipped on their own to make those judgements; they can't evaluate the model, or even find it very easy to decide do we want a model to be 90 per cent accurate before we make policy or 60 per cent accurate. So actually what has happened in climate change is that national governments, I'm talking mostly I suppose here from the UK perspective, set up advisory panels where scientists really could assess the evidence and inform policy advisors.

I've been part of one or two of those exercises, but much more important on a global scale was the establishment of the IPCC which was commissioned by the United Nations to assess the state of climate change knowledge including how good the models are. And I would argue that most of the policy decisions that have been developed around climate change have used the IPCC's assessments, to greater or lesser extent, to inform the policy process. The IPCC has been extremely influential as a mechanism that sits in between science and policy.

The Tyndall Centre and the IPCC

Q The way that the Tyndall Centre would relate to the IPCC is that there are some individuals that work with the IPCC?
A Yeah, the Tyndall Centre came along in 2000 and had a much wider remit than just looking at models and data. Our remit was to undertake

interdisciplinary research. We're looking at the scale of the problem and the range of options that might tackle the problem and that covers a huge intellectual territory. The Tyndall Centre is recognised by the IPCC as an expert institution and so we can comment on the IPCC process and draft documents. We can nominate individual scientists to attend IPCC Plenaries as observers. What we can't do is put forward the names of individual scientists to become a member of the IPCC because that has to come from government. So, in the UK, it's Defra who has to nominate scientists. We can say to Defra 'well, we've got three or four guys here who are really good experts on adaptation, mitigation and all that. You might want to put them forward.' And we have a number of, six or eight, people within the Centre who are actually involved in writing some of the present (4th) assessment. But our relationship has to go through government because the IPCC is an inter-governmental organisation and it is governments who have the power to nominate experts.

Q There hasn't been any friction in that process has there? Has it been more or less amenable?
A Yes. I've got my personal views about the IPCC. But as an institution, as the Tyndall Centre there's no problem. I think the only issue, at the institutional level, I would say is that scientists who are invited to play a part in the IPCC are invited to do so without any remuneration for their time and effort. They have their expenses paid to travel to a meeting, but they don't get any remuneration for their time. In effect, what that means is that the scientists' employers are subsidising the IPCC process because my colleagues' time is offered free. And sometimes that means that certain scientists will say 'well, I'm not going to accept, I'm busy'.

And then an individual scientist's decision to participate or not is a cost-benefit question. Is the time I'm personally putting in, do I get more out of it than I put in? And the benefits are less tangible benefits. You don't get any remuneration but you do get access to networks, you meet other scientists from around the world, you perhaps get access to new information that you might not have known about. It certainly doesn't do your promotional aspects any harm. It is seen in the academic system as a measure of esteem. But they're intangible benefits, they're not immediate financial benefits and scientists will certainly view the balance differently. For example, I was very heavily involved in the Third Assessment of the IPCC a few years ago, but I've resisted all invitations to be involved in this Fourth Assessment because I reached a different decision this time. It wasn't in my interest to do this.

Influences on the Tyndall Centre's research agenda and funding

Q You mentioned something about academia. The Centre encompasses a network of individuals, 200 or so researchers, and they all have different ties and so forth. What's the influence of academia in the work that gets done in the Centre?

A There are a number of different influences on that agenda. One of which has to be the influences that come from the academic system. There are more powerful influences, I suspect, that come from our dialogue and discussion with stakeholder interests including business, government and civil society. There are also influences on our agenda that come from our funders, the research councils. In the end, as director, me and my management team have a large say about exactly what research we do in the end. We're not acting as a consultant to a client. We're not acting as a contractor to government. It's not as though we tendered for a particular piece of research for government to fund and then said 'we want so much money to do this piece of research'. The Tyndall Centre is funded primarily by the research councils and the research councils, in this case, have given us fairly good autonomy in how we set our research agenda. They said 'we'd give you the £10 million, we liked your vision, how you would operate, the sort of experts you'd bring into play, the sort of large-scale questions that you would be looking at'. But how we put flesh on that and allocate money to do this bit of research and not that bit of research was largely left up to us.

So, the Tyndall Centre, as with most research council investments, is quite different from a consultancy or from tendered contract research. But there are influences clearly. You asked a question about the academic system … I suppose the influences from the academic system come largely through the way in which individual scientists, academics, most of us are employed by universities. How we perceive the career structure that we operate within and what measures of success or esteem that the system recognises as important.

The way in which that works in the university system is the research assessment exercise, the RAE. For our employers, our universities, getting a good RAE grade is of fundamental importance. So that cascades down to individual departments and to individual academics. We all are working within this RAE game, where classically, still, the benchmarks are peer-reviewed journal papers in leading international outlets, for example *Science* and *Nature* in our area. And that influences the way in which individuals and academics think of their own priorities.

For the Tyndall Centre this introduces some interesting tension because, on the one hand, we've been asked to work on the interface with society to look at options, solutions to climate change, and that often means we come into serious contact with people in government, business, civil society and the media. And the expectation is that we are actually doing some kind of applied, user-oriented research. It's not easy always to translate that research into peer-reviewed papers for *Science* and *Nature*. So, there are some tensions here that clearly exist. The academic system through the RAE is expecting us as scientists to do a certain type of research. The expectation elsewhere in society and in the research councils is that the Centre should really be focusing on user-oriented, solutions-driven research.

Q Could you possibly tell me, if this is something that you're allowed to talk about or you want to talk about … but with the Phase 2 proposal,[47] are any of these tensions going to be addressed? Or what are going to be the major things that are addressed in Phase 2?

A We're still in the middle of the process of negotiating a further five-year extension for our funding base and there are many checks and balances that seem to exist in this decision process. There are different expectations here. So, the research councils, through whom the public money is channelled, have their own expectations and criteria. Again, clearly there are demands that come from our university system although, in fact, the only demand that comes from the university system is 'get the money' because it's a major measure of success if the university can bring in a big, fat research contract. They're not too worried about what we do, they're just interested in securing it.

But then society at large has certain expectations as well. So, certain agencies in government would expect the Tyndall Centre to do research that meets some of their internal needs for policy purposes. And in the end it's a matter of judgement how we balance these interests. The decision process in the research councils is the most critical one because they determine the level of core funding. And in preparing for the decision process there were various steps in that chain. The critical one is what's called the 'moderating panel' comprising ten experts, supposedly independent, that the research councils appoint to scrutinise our proposal. And these experts could say 'this is fine', they can say 'this is rubbish'. They can also say 'it's got strengths and weaknesses' and recommend that we change it.

It's very interesting being in the middle of this process right now to observe what a mixture of influences, factors, motives shape the way a committee reaches a decision. And we haven't been given an easy time; I mean it's no secret that our original proposal was turned down [in January 2005] and we've now been asked to resubmit a different version. And the interesting thing is that the expectation from the wider network, the wider community of people who know anything about the Tyndall Centre think 'how on earth can they [the research councils] turn down funding when climate change remains such a high priority' certainly

47 This interview took place in the middle of an extended 18-month process in which the research councils evaluated the Tyndall Centre and decided on what level of funding to award the Centre in Phase 2 (2005–2010) of its existence.

on the UK policy agenda. This is the year 2005, Tony Blair chairs the G8, the UK chairs the EU, and he's got climate change as one of his top two concerns. So, how can the UK Government be saying that 'this is so, so important' and yet UK science will not fund a major national research capacity? People make that fairly simple observation.

But of course it's not as simple as that because of the way in which decisions are delegated through the system. So this moderating panel is looking at the criteria. They're looking at scientific excellence, they're looking at whether we are meeting user needs, they're looking at what we've done with the money, they're looking at whether we've got the right management structure given the organisation is spread across the country. And then there are personal factors. Individuals [on the panel] like some people and don't like others. These personal factors influence the way that decisions get made.

Q As I understand it, the moderating panel is a composition of other academics, right?
A They're largely other academics, although there are one or two people who might represent non-academic interests, i.e. there's one senior person from business, one quite senior person from an NGO.

Q Because I'm almost wondering if it's a situation where they're looking at the Centre's funding from a personal standpoint; where they say 'if we allocate this money to this organisation then that money might be taken away from my department'. Do you get the impression that it comes down to a personal level or is enough money spread around so that isn't a factor?
A Yes, I think that perspective is there in the minds of other academics, but I don't think it's unique to the Tyndall Centre. I think any peer review system that relies on academics to evaluate other academics – whether it's funding, whether it's scientific journals – there is clearly … let's call it 'jealously' or 'vested interest'. Scientists are as human as anyone else and we all know there are certain things that influence us to a greater or lesser extent. Clearly, one would argue the more professional the individual concerned is they put those things, those thoughts, right into the background and say 'that's not really what we're judging on, we're judging purely on the merit'. And I think, for the most part, some of the people are professionally driven. But you can't pretend that the system doesn't have that weakness. Probably it's a bit like Winston Churchill's comment that 'democracy is the worst of all government systems except for all the others'. And peer-review is probably the worst possible system for conducting science except for all the others.

Tyndall Centre 'messages'

Q Leaving that process aside for the moment, in terms of the vision that's being rearticulated now, is there going to be a change in the way that the topic of climate change is presented by the Centre to the public?

A   Well, I'm not certain that I can be so bold as to say that the Tyndall Centre
has radically altered the way in which climate change is presented to the
public. We do quite a lot of engagement through public lectures, public
speaking, advising, popular articles, or, indirectly by talking with the media,
which is quite a powerful filter. So, the Tyndall Centre certainly sees itself as
an actor in the ways in which climate change is communicated and perceived
within society. And it wasn't just something that we fell into by accident, it's
not simply something that we do reactively. One of our three core objectives
as a Centre is to promote informed debate and dialogue across society about
the options with respect to climate change. So it's built into the way in which
we operate, I think. In some tangible or intangible ways we've probably altered
some of the terms of the debate or some of the perceptions of the debate in the
last few years. Inevitably if you're active as an agent in the system, the system
is different than if you weren't active and present in the system.

Trying to pin it down to specifics though becomes much more problematic.
Because then you're looking at the whole science and society interface and what
influences the ways in which society perceives an issue. Again, clearly, there are
lots of influences on that. What I would say is that one of our key messages –
certainly this is a key message that I've championed for many years – is related to
adaptation to climate change. Any decisions that are made by organisations in
society, government, private sector, civil society that are sensitive to climate need
to make sure that those decisions are based on the perspective that future climate
will change, rather than based on the assumption that the climate of the past
gives you all the information that you need about the future. It's an important
message for public and private sector organisations who are making investment
decisions, design decisions. We have to design the infrastructure, the institutions,
the management systems across a whole variety of sectors, we've got to design
society to the climate of the future and not the climate of the past. That's one key
message that I use on many occasions.

I think another key message is to try to put across the idea that climate change
is not a problem waiting a solution to be discovered by science ... that in a sense,
climate change isn't a problem. Climate change is simply a consequence or
manifestation of a globalising world which has a certain energy system – fossil
carbon based. Climate is a changing boundary condition of society. Climate itself
is morally neutral. There is no such thing as a 'bad' climate or a 'good' climate or
a 'dangerous' climate or a 'safe' climate. It all depends on how we as humans
decide to live. There are flourishing societies in Iceland, there are flourishing
societies in Saudi Arabia. The climates of those two countries are hugely different.
People can live very sustainably in both climates if society is appropriately
designed.

And that's the same about climate change. It's not as though it's some absolute
moral problem of danger, it is simply challenging society about how we organise
ourselves. That's certainly my personal take on some of the Centre's messages.
Although what you'll find, of course, is that I'm not speaking on behalf of the

Centre, this is not the corporate message. I am speaking here as Mike Hulme and what you'll find in this organisation are quite different views about climate change and how it should be communicated to the public.

Q   In terms of something that might be a uniting principle or belief amongst the researchers here, do you get the impression that not everyone is just chasing funding (if that's what it seemed like in Phase 1)? I mean do you think there was something more?

A   Well I think again, let's just stick to the fact that any institution is made up of human beings and what motivates us varies. I think there are quite a number of our researchers or scientists who believe quite passionately in the need for society to take climate change seriously and who genuinely think that the Tyndall Centre is the right way for science to make its contribution. I think there are other people, maybe people who are further on in their career who are much more pragmatic, which is another way of saying less idealist. And it probably comes back to the question about the academic system and the pressures on academics within that system. Funding is funding, status is status, esteem is esteem, approval by the RAE is approval by the RAE and they'll do whatever it takes to get those successes. Maybe that's too cynical, I'm not sure.

So, I think you'll find in any organisation a range of motives. But I would say that many of my staff do genuinely think that the Tyndall Centre is where they want to be at the moment. And for that reason they think that the Tyndall Centre is the right way to organise science's contribution to policy solutions ... rather than just having a specialist department in one university doing some quite narrow writing on economics, or another department somewhere else doing some quite narrowly defined project on hydrological modelling. They genuinely believe that trying to bring together different disciplines to open up new perspectives, to illuminate some of the difficult choices that society faces is the right way to do it.

Q   Was the focus on stakeholder interaction and on public debate something that was initially in the research councils' idea for the new climate change research institute or was that something that came out of your proposal and they said 'wow, that's interesting. We like that, we'll go with that'? Was that something that differentiated you from the other proposals, or did you even know about what the other designs were going to be?

A   If you haven't seen it, I can give you the original specification from the research councils in 1999. It's only a short document but it shows what the research councils thought they were creating. And then clearly, we had to put flesh on that and operationalise and interpret it in a certain way. I think all bids had to respond to the user interface because that was part of the original specification, but there are different ways, of course, of doing that. And we've tried to do that in a certain way by finding partnerships, by opening up dialogues at least to some extent and by being responsive to the expectations of potential users and the rest of society.

How far did we get that down road, or have gone, or should go, is again an interesting debating point, because if you ultimately go down that road you end up becoming consultants. So, if you get too close to BP or Defra or Greenpeace and listen to them too carefully and respond to them too precisely, all you're then doing is doing things that they're dictating. And, as we know with a lot of consultancy work, quite often consultancy work is about providing answers to questions which the client already 'knows' the answer to, rather than actually opening up in a more genuine independent spirit of enquiry; i.e., generating an answer that no one actually imagined at the beginning.

So, this is an issue again that we've debated internally in the Centre, where on that continuum do we want to be. In a very crude sense you've got one point which is the ivory tower, blue sky. An academic gets a grant, goes away and forgets about the outside world. The other end of the spectrum is a consultancy, pure consultancy, where you take the money and do what the client wants. Where on that spectrum should the Tyndall Centre be? It clearly can't be at either end, but should it be somewhere in the middle ... there's no right answer to that, but it does influence the way in which we do our science, the way in which we communicate our messages.

Certainly it influences the expectations that other agents in society have about the Tyndall Centre. We've been criticised by some agencies for not being responsive enough. Have we been criticised for being too responsive? Yeah, I think implicitly we have been criticised for being too responsive and probably this comes back to this thing about the RAE. The RAE measures the fairly classical, conventional academic criteria at that end of the spectrum. And if we don't do enough of this, then we'll get criticised: 'where are your papers in *Nature?*'

Uncertainty, decision-making and the framing of climate change

Q The second objective that you mentioned was promoting a dialogue about climate change. One of the hinges that I think gets the debate bogged down is all this talk about uncertainty and so forth. How is that particular issue approached?

A My take on this is that climate change emerged very much from the physical science community as a scientific curiosity, particularly as the first work was done in the 1970s. It ramped up during the 1980s before it broke through the public realm. There's the physical climate science community, meteorologists fundamentally, who moved into climate change. And meteorologists are all about forecasting; that was their traditional role in society to forecast the weather for tomorrow. And uncertainty in the forecast, getting the forecast wrong or right, was a key dimension of all meteorological forecasting work. And people always complain that their forecasts are inevitably wrong, there's always uncertainty.

Anyway, some of the ways in which we present uncertainty in climate change are inherited from where the issue started off in meteorology and weather forecasting. And that has been a problem for climate change in society because it has given too much focus to knowing how physical climate systems will evolve, as though we can only make decisions about the future when we have certain information. Whereas actually when you think across most decisions that humans and organisations make about the future, there are huge uncertainties that bear on those decisions ... the interest rate set by the central banks, the price of oil, or at a personal level what my health status is going to be in my retirement, the rate of return on an investment. And who is out there in the world of science saying 'well we've got a model that can tell you, over the next fifty years, the rate of return on an investment and this is the 10 per cent range'.

So, the way in which we make decisions as a society and as individuals always confronts massive uncertainties and yet we have to make decisions based on the best knowledge ... or guesses, or prejudices, or values. My view is that many are convinced that climate change is just too uncertain to take it seriously. This is because climate change is still couched in terms of predictions that scientists make about particular physical systems and uncertain predictions implies poor science, inheriting the language from weather forecasting where uncertainty is everything.

There have been a few people, including me, who want to frame climate change in a completely different way. It focuses on decisions made today which have future implications. How do we make decisions and what information is needed, what assumptions are made, what priorities are implied by different decisions? We look at a decision and say ... 'well, what information do we need about future climate? Do we require that information to be 100 per cent accurate? Do we want uncertainty to be expressed as a probability – you know there's a 90 per cent chance that something like that will occur, or a 50 per cent chance?' So, if you focus the issue around decision-making and how people make decisions and what influences come to bear on those decisions, then changes in future climate can take its place alongside a whole range of other things that will change as well.

It's a very strange situation I think, but I can see the reasons why we have become so fixated on climate predictions. Like I said before, because of the intellectual origins of climate change we unfortunately inherited the expectation that climate change [research] is all about narrowing uncertainty and [climate change] only becomes relevant once the uncertainty is reduced to a certain level.

Q Now, you said that you and a few other individuals have tried to cultivate the decision-focused frame. Is that expressed in your research and in the way that you talk about climate change to policy-makers and the public?

A There are some formal ways in which our research could take that perspective, whether it's in relation to adaptation or mitigation. It's quite difficult in some policy or public debates because it is then about changing mindsets which have been set by the way the debate has evolved. And the media often keep bringing it back to the rather sterile 'is this happening because of climate change, or is not; or is it certain, or is it not'. Changing the frame of reference for the media is quite hard and often for policy advisors as well who have inherited a certain framing of climate change. My experience, at least in talking with policy people, is that for them it still too often comes back simply to narrowing uncertainties rather than trying to reframe climate change using a different starting point.

And we're not helped by some of the rather simplistic rhetoric that comes out of campaigning organisations who like to up the ante and say 'this is going to happen, that's going to happen, because of climate change'. They expose themselves to the obvious criticism that the science is too uncertain to make such claims. You know they say that the Gulf Stream is going to collapse or that we're going to lose half of our species, or that Arctic ice will disappear. If you look at the science there's a lot of uncertainty about where we are. Yet they continue to hammer out catastrophic depictions of climate change and that all these scientists are predicting global catastrophe. They are distorting the debate. So, the debate gets dragged back again to 'well, is this scientist correct, is this prediction correct?' rather than thinking 'what are the decisions that society faces upon which future climate might have a bearing?'

This latter approach is trying to use what we know about climate to reach robust decisions that are insensitive to the uncertainties that we know about. We're not going to get rid of uncertainties about future climate with just another big science push. You know the generals in the First World War used to say to their troops 'one more big push and we'll break through the deadlock in the Western trenches; one more big push will do it.' There's one more big push and another million lives are lost! It's the same thing sometimes with climate scientists: 'If we put a really big science programme together, £40 million, then we will crack the uncertainty problem once and for all'. I don't see climate change like that.

Q   Okay, this will be my last question. How is it that you can start that new framing? I mean you have quite a lot of baggage here.
A   There are people who wouldn't share my approach to the problem ... there are others who might disagree with me or just don't have the flexibility to be able to reframe things. So, I can't claim that this is the distinctive Tyndall Centre message. All I can say is that this is partly my own intellectual journey and I know some others in the Tyndall Centre share those views with me. It seems to me that the Centre should be promoting this more realistic approach, opening up the issue in this sort of way because it is much more creative. And it also takes us away from the rather futile debate that gets rehearsed time and time again about uncertainty: 'is climate change happening and how accurate are the models?' Instead, it moves the debate onto what really is, I think, the legitimate debating point in climate change which is about values.

There are legitimate differences in our societies about how much weight we should give to climate change. But the differences are not because we believe *this* model and the other people believe *that* model. The differences are because we value things differently, our political visions are different, we weight environmental regulation and deregulation differently … and that takes us back into our political vision of good government.

How we value the preservation of ecosystems and biodiversity, for example, will vary legitimately between people and it's quite appropriate to have debates about those things. But continually thinking that the climate change debate is about 'is that model correct, is this model correct, or is the hockey stick curve' – I don't know if you know about that – 'is it this shape, or this shape, or this shape?' is unhelpful. Okay, it's fine for scientists to debate these points, but if the public think that this is what climate change is all about then they completely misunderstand the nature of climate change in relation to modern society. We should generally be using climate change to make explicit that we value things differently. And it seems to me that this is exactly what a democracy is there to accomplish; to tease these things out and to reach decisions through democratic institutions, not through bullying people with scientific claims.

Q   But can't you as Executive Director say 'this is the vision that we're going for, this is what we want to realise'? I know it's forcing a vision on other people, but maybe they need to be coaxed a little bit or they need to be prodded because maybe they are inflexible with their thinking.

A   That's an interesting challenge, I suppose. It's a leadership challenge. Maybe that is the role of a leader, of a visionary. I don't necessarily cast myself in that role. I probably tend to see my role, for good or bad, more about making sure the institution works, making sure that it's effective, that it's well-managed, because there are a lot of very creative and smart people in the organisation … That's interesting though, actually putting that challenge to me. Executive Directors tend to see themselves more as a managerial …

Q   But in terms of motivating your 'employees', if you want to call it that …

A   I mean, I certainly don't stand out there and bang the drum. I certainly have some individuals that I am working with where we discuss these things and certainly I've followed through and reflect some of those elements in my public lectures, my public talks. I give quite a lot of lectures to school kids usually and to the more general public. I tend always try to present these issues more in the language of scientific uncertainties and value conflicts.

17 The appliance of science

14 March 2007

Review essay written for *The Guardian* newspaper[48]

> Politicians and the public look to scientists to explain the causes of climate change and whether it can be tackled – and they are queuing up to deliver. But, asks Mike Hulme, are we being given the whole picture?

Climate change is happening, but it appears that science is split on what to do about it. One of the central reasons why there is disagreement about how to tackle climate change is because we have different conceptions of what science is, and with what authority it speaks – in other words, how scientific 'knowledge' interacts with those other realms of understanding brought to us by politics, ethics and spirituality.

Two scientists – one a climate physicist, the other a biologist – have written a book arguing that the warming currently observed around the world is a function of a 1,500-year 'unstoppable' cycle in solar energy.[49] The central thesis is linked to evidence that most people would recognise as being generated by science. But is this book really about science?

It is written as a scientific text, with citations to peer-reviewed articles, deference to numbers, and adoption of technical terms. A precis of the argument put forward in the book by Fred Singer, an outspoken critic of the idea that humans are warming the planet, and Dennis Avery is that a well-established, 1,500-year cycle in the Earth's climate can explain most of the global warming observed in the last 100 years (0.7°C), that this cycle is in some way linked to fluctuations in solar energy, and because there is nothing humans can do to affect

48 Hulme, M. (2007) The appliance of science in *The Guardian*, 14 March, G2, p.9 http://www.guardian.co.uk/society/2007/mar/14/scienceofclimatechange.climatechange. A slightly amended version of this review essay also appeared in *Weather* (2007) 62(9), 243–244 as 'Understanding climate change: the power and the limit of science'.

49 Singer, F.S. and Avery, D.T. (2007) *Unstoppable climate change every 1500 years* Rowman & Littlefield, Latham, USA.

the sun we should simply figure out how to live with this cycle. We are currently on the upswing, they say, warming out of the Little Ice Age, but in a few hundred years will be back on the downswing. Efforts to slow down the current warming by reducing emissions of greenhouse gases are at best irrelevant, or at worst damaging for our future development and welfare.

This, of course, is not what the fourth assessment report of the UN Intergovernmental Panel on Climate Change (IPCC) said a few weeks ago. The report from its climate science working group concluded that it is likely that most of the warming of the last 50 years has been caused by rising greenhouse gas concentrations and that, depending on our actions now to slow the growth of emissions, warming by 2100 will probably be between about 1.5°C and 6°C. The upper end of this range is almost an order of magnitude larger than the warming that Singer and Avery suggest is caused by the 1,500-year cycle. So is this a fight between scientific truth and error? This seems to be how Singer and Avery would like to present it – 'science is the process of developing theories and testing them against observations until they are proven true or false'.

Means of inquiry

At one level, it is as simple as this. Science as a means of inquiry into how the world works has been so successful because it has developed a series of principles, methods and techniques for being able to make such judgements. For example, we now understand the major transmission routes for HIV/Aids, that smoking injures health, and that wearing seat belts saves lives. And so it is with climate change. Increasing the concentration of greenhouse gases in the atmosphere warms the planet and sets in motion changes to the way the weather is delivered to us, wherever we are. Science has worked hard over a hundred years to establish this knowledge. And new books such as Singer and Avery's, or opinion pieces in *The Daily Mail*, do not alter it.

So far so good. Deploying the machinery of scientific method allows us to filter out hypotheses – such as those presented by Singer and Avery – as being plain wrong. But there are two other characteristics of science that are also important when it comes to deploying its knowledge for the benefit of public policy and society: that scientific knowledge is always provisional knowledge, and that it can be modified through its interaction with society.

That science is an unfolding process of discovery is fairly self-evident. The more we seem to know, the more questions we seem to need answering. Some avenues of scientific inquiry may close off, but many new ones open up. We know a lot more about climate change now than 17 years ago when the first IPCC scientific assessment was published. And no doubt in another 17 years our knowledge of how the climate system works and the impact that humans have made on it will be significantly different to today.

Yet it is important that on big questions such as climate change scientists make an assessment of what they know at key moments when policy or other collective decisions need to be made. Today is such a time. But our portrayal of the risks of climate change will always be provisional, subject to change as our understanding advances. Having challenges to this unfolding process of discovery is essential for science to thrive, as long as those challenges play by the methodological rule book that science has painstakingly written over many generations of experience.

The other important characteristic of scientific knowledge – its openness to change as it rubs up against society – is rather harder to handle. Philosophers and practitioners of science have identified this particular mode of scientific activity as one that occurs where the stakes are high, uncertainties large and decisions urgent, and where values are embedded in the way science is done and spoken. It has been labelled 'post-normal' science. Climate change seems to fall in this category. Disputes in post-normal science focus as often on the process of science – who gets funded, who evaluates quality, who has the ear of policy – as on the facts of science.

So this book from Singer and Avery can be understood in a different way: as a challenge to the process of climate change science, or to the values they believe to be implicit in the science, rather than as a direct challenge to scientific knowledge. In this reading, Singer and Avery are using apparently scientific arguments – about 1,500 year cycles, about the loss of species, about sea-level rise – to further their deeper (yet unexpressed) values and beliefs. Too often with climate change, genuine and necessary debates about these wider social values – do we have confidence in technology; do we believe in collective action over private enterprise; do we believe we carry obligations to people invisible to us in geography and time? – masquerade as disputes about scientific truth and error.

We need this perspective of post-normal science if we are going to make sense of books such as Singer and Avery's. Or indeed, if we are to make sense of polar opposites such as James Lovelock's recent contribution *The Revenge of Gaia*, in which he extends climate science to reach the conclusion that the collapse of civilisation is no more than a couple of generations away. The danger of a 'normal' reading of science is that it assumes science can first find truth, then speak truth to power, and that truth-based policy will then follow. Singer has this view of science, as do some of his more outspoken campaigning critics such as Mark Lynas. That is why their exchanges often reduce to ones about scientific truth rather than about values, perspectives and political preferences. If the battle of science is won, then the war of values will be won.

If only climate change were such a phenomenon and if only science held such an ascendancy over our personal, social and political life and decisions. In fact, in order to make progress about how we manage climate change we have to take science off centre stage. This is not a comfortable thing to say – either to those scientists who still hold an uncritical view of their privileged enterprise and who relish the status society affords them, or to politicians whose instinct is so often to hide behind the experts when confronted by difficult and genuine policy alternatives.

Two years ago, Tony Blair announced the large, government-backed international climate change conference in Exeter by asking for the conference scientists to 'identify what level of greenhouse gases in the atmosphere is self-evidently too much'. This is the wrong question to ask of science. Self-evidently dangerous climate change will not emerge from a normal scientific process of truth seeking, although science will gain some insights into the question if it recognises the socially contingent dimensions of a post-normal science. But to proffer such insights, scientists – and politicians – must trade (normal) truth for influence. If scientists want to remain listened to, to bear influence on policy, they must recognise the social limits of their truth seeking and reveal fully the values and beliefs they bring to their scientific activity.

Chink of weakness

Lack of such reflective transparency is the problem with 'unstoppable global warming', and with some other scientific commentators on climate change. Such a perspective also opens a chink of weakness in the authority of the latest IPCC science findings.

What matters about climate change is not whether we can predict the future with some desired level of certainty and accuracy; it is whether we have sufficient foresight, supported by wisdom, to allow our perspective about the future, and our responsibility for it, to be altered. All of us alive today have a stake in the future, and so we should all play a role in generating sufficient, inclusive and imposing knowledge about the future. Climate change is too important to be left to scientists – least of all the normal ones.

18 Geographical work at the boundaries of climate change

26 November 2007

Essay written in the 'Boundary Crossings' series of the journal *Transactions of the Institute of British Geographers*[50]

Introduction

The relationship between climate and society has been dynamic throughout human history and pre-history, a relationship that has been variously elemental, creative and fearful. The relationship has now taken a more intimate turn. Human actions, globally aggregated, are changing the composition of the atmosphere which alters the functioning of the climate system. Future climates will not be like past climates. We have often worried about this possibility and now the knowledge claims of science have offered new reasons to be concerned. Humanity is now firmly embedded within the functioning of the climate system, whilst at the same time the *idea* of climate change is penetrating and changing society in novel ways. The past (through historic emissions of greenhouse gases) and the future (through descriptions of climates to come) are interacting in new ways to provide a novel motor for cultural change. This is all happening under the symbolism of global warming.

Yet the construction of narratives around global warming remain strongly tied to roots within the natural sciences, to expectations of improving 'predictions' and to a problem-solution policy framing which claims both global reach and universal authority. The European Union policy goal of restricting global warming to no more than 2°C above nineteenth-century global temperature is seemingly powerful, yet fundamentally fragile: powerful because it traces its lineage to positivist and predictive sciences; fragile because it is largely a construction of elite Western minds. This constructed policy goal is unlikely to be one around which the world will be re-engineered willingly. Neither positivist science nor Western (neo-)liberalism seem likely to retain global hegemony.

50 Hulme, M. (2008) Geographical work at the boundaries of climate change *Transactions of the Institute of British Geographers* 33(1), 5–11. ©2007 The Author. Journal compilation © Royal Geographical Society (with The Institute of British Geographers) 2007.

The emergent phenomenon of climate change – understood here simultaneously as physical transformation and cultural object, as a mutating hybrid entity in which the strained lines between the natural and the cultural are dissolving – therefore needs a new examination. And this re-examination must have a different starting point from that arrived at in the twentieth century. Those origins were to be found in the scientific disciplines of the natural sciences and in the institutional process of the Intergovernmental Panel on Climate Change (IPCC), a process whose outcomes rapidly came to dominate climate change discourse. They still do. Instead, our re-examination of climate change today needs to start with contributions from the interpretative humanities and social sciences, married to a critical reading of the natural sciences, and informed by a spatially contingent view of knowledge (e.g. Pettenger, 2007).

The questions that should guide this re-examination include these: What does climate mean to different people and to diverse cultures? Which of these meanings are threatened by climate change and which can co-evolve with a changing climate? How robust is our putative knowledge of future climate? What language is used to portray climatic risks? Is climate change really a collective action problem? Who gains from driving forward ideas of global climate governance? And, in the end, what is our vision of the global future? Who speaks for the twenty-second century? The answers to many of these questions are answers to which geographers should contribute. Shy of intellectually owning climate because of the shame of past determinist ideologies (see Skaggs, 2004), the society of geographers needs now to re-embrace the idea of climate.

I am writing this essay as a geographer who has become detached from the discipline over a 20-year career working at the heart of the climate change knowledge community, close to the circuits of policy. Yet I am troubled by what I have observed happen to climate change. I am increasingly convinced that making human sense of climate change needs the distinctive intuition and skills of the geographer. These intuitions include long familiarity with working at the boundaries between nature and culture (e.g. Petts *et al.*, 2008), a tradition of understanding the subtleties of how knowledge, power and scale are inseparable (e.g. Cox, 1997), and the more recent unveiling (e.g. Livingstone, 2003; Powell, 2007) that the geography of science is every bit as important as its history. We need new ways of thinking about and understanding the hybrid phenomenon of climate change. Geographers have a unique role to play in this task.

The essay will suggest three fluid boundaries surrounding climate change at which geographical work could fruitfully be located. The first of these is the important project to reclaim climate from the natural sciences and to treat it unambiguously as a manifestation of both Nature and Culture, to assert that the idea of climate can only be understood when its physical dimensions are allowed to be interpreted by their cultural meanings. The second project, contingent on the first, is to reveal that discourses about global climate change have to be re-invented as discourses about local weather and about the relationships between weather and local physical objects and cultural practises. Climate change knowledge and meaning travels uncomfortably across scales and needs constant re-interpretation as it is applied in different spatial contexts. The third project is to examine critically the knowledge about climate change captured by the various assessments of the IPCC and to understand the various ways in which this knowledge is situated. The IPCC assessments exert considerable hegemony over climate change policy debates and yet insights from the geography of science can tell us a lot about how (un)stable and particular such knowledge is.

Reconnecting culture with climate

The framing of climate change that remains dominant in contemporary scientific and public policy discourse was shaped crucially during the period from 1985 to 1992. The WMO/UNEP/ICSU Villach Conference of November 1985 established the hegemony of the natural sciences in the way climate change would subsequently be presented to the policy world. This framing led directly to the establishment in 1988 of the IPCC under the auspices of the WMO and UNEP (Miller, 2004). Through its first Assessment Report of 1990 the IPCC staked out the contours of climate change – almost trade-marked *Climate Change*™ one might say – which were to dominate the next two decades. It certainly had an imposing effect with respect to the inter-governmental negotiations which led in 1992 to the signing of the UN Framework Convention on Climate Change. The Convention, supplemented by the Kyoto Protocol in 1997, remains today the benchmark for all climate policy discourse, a benchmark bolstered by the knowledge claims in subsequent reports from the IPCC in 1995, 2001 and 2007.

The three key elements of this framing were: a globalised atmosphere – 'an ontologically unitary whole' (Miller, 2004) – which offered to the world a single depository for greenhouse gas emissions and which opened the way for predictive climate modelling; the goal of a stabilised global climate as the centrepiece of policy; and the institutionalising of mitigation and adaptation as co-dependents in future global climate policy regimes. Insights from the social sciences, and from geography in particular, were notably absent from this early framing. The considered contributions from a social science perspective – most notably Rayner and Malone (1998) and Miller and Edwards (2001) – came along too late to re-mould the framework despite attempts to do so (e.g. Sarewitz and Pielke, 2000).

The consequences of this history remain with us today. Arguably (e.g. Sarewitz, 2004), they are one of the reasons why progress towards meeting stated climate change policy goals is so tortuous. Climate is defined in purely physical terms, constructed from meteorological observations, predicted inside the software of Earth system science models and governed (or not) through multi-lateral agreements and institutions. What is sought to be stabilised is a quantity – global temperature, or its proxy carbon dioxide concentration – a quantity wholly disembodied from its multiple and contradictory cultural meanings. Future climates are predicted and, whether these predictions are read rhetorically or literally, they depend tenuously – at best – on ideas and possibilities of future cultural change.

Yet there is another, orthogonal reading of climate which has been suppressed in this contemporary framing of climate change. Masterfully narrated by Clarence Glacken in his 1967 *Traces on the Rhodian Shore*, climate can also be understood as an imaginative idea, an idea constructed and endowed with meaning and value through cultural activity. A few more recent voices have been raised in praise of such a reading, the voices of cultural historians, historians of science and anthropologists; thus books about climate and culture by Meyer (2000), Strauss and Orlove (2003), Boia (2005), Cruikshank (2005), Fine (2007) and Golinksi (2007). Registers of climate can be read in memory, behaviour, text and identity as much as they can be measured through meteorology, as Golinski shows in his enlightened survey of eighteenth-century British attitudes to weather and climate: 'this … view of the climate was bound up with the sense of British national identity' (2007: 57). Of these voices, however, few have been geographers.

So why do we persist with this dualistic account of climate as non-overlapping physical and cultural entities? Are we content with this 'purification' of climate, a purification that Latour (1993) claims is constitutive of modernism? My claim in this essay is that the answer is 'no' and that geographical work can help bring these two orthogonal readings of climate into a more creative alignment. This does not merely imply siren calls for greater inter-disciplinarity in climate change research institutions or funding. It implies something deeper than this. The subject to be studied – climate – needs to be re-framed through negotiating a different ontological and epistemological structuring of climate knowledge (cf. Petts *et al.*, 2008). This intellectual re-framing of climate has to take place *before* one considers the methods, tools and institutions which would then need to be deployed.

If we are to show the multitudinous ways in which climate and culture are connected in places that have a history and in places that have a future then William Meyer's (2000) cultural account of Americans and their weather needs re-writing for multiple ethnicities, networks and nation-states. Lucien Boia's (2005) imaginative history of the idea of climate needs its geographical sibling. Many more of the types of case studies reported in Strauss and Orlove (2003) need completing. By critiquing the geopolitics of the anthropocene, through tackling the triumvirate of globalisation, empire and environment, Simon Dalby (2007) shows glimpses of how a new holistic and critical geography of the natural and the cultural can yield novelty.

Moving climate across scales

The sterility of a purely physical reading of climate can be illustrated if we consider the movement of statistically constructed climates across scales. The predominant way of capturing the physicality of climate is through meteorological measurement.[51] By standardising such measurements and then by circulating them through centralised bureaucracies it became possible, first, to quantify (local) weather and subsequently to construct statistical (aggregated) climates. The story of this 'domestication' (Rayner, 2003) of climate through the eighteenth and nineteenth centuries is well told by, respectively, Golinski (2007) and Anderson (2005). More recently in the twentieth century, these standardised registers of weather have enabled an abstracted 'global climate' – indexed via surface temperature – to be constructed, opening the way for model-based predictions of climate change and fuelling a hubristic discourse about the 'control' of future climate (compare Crutzen, 2006 with Fleming, 2006).

It is important to notice what happens in this circuit of transportation. Weather is first captured locally and quantified, then transported and aggregated into regional and global indicators. These indicators are abstracted and simulated in models before being delivered back to their starting places (locales) in new predictive and sterilised forms. 'Digitised' weather for virtual places can even be conjured from these models using stochastic weather generators. Through this circuitry, weather – and its collective noun climate – becomes detached from its original human and cultural setting. A rainstorm which offers an African farmer the visceral experience of wind, dust, thunder, lightening, rain – and all the ensuing social, cultural and economic signifiers of these phenomena – is reduced to a number, say 17.8 mm. This number is propagated into the globalised and universalising machinery of meteorological and scientific institutions and assessments where it loses its identity.

There is of course much more to this story, but the point of telling it this way is to show that climates do not travel well between scales: the essential loading of climate with culture – what climate means for people and places and the relationships between people and places over time – is completely lost through such purifying practices. This begins to explain one of the paradoxes of the current framing of climate change. On the one hand the physicality of weather is being increasingly influenced by human practises on a global scale – yes, temperatures *are* rising. Yet the very construction of these universalised diagnostic indicators of change strips them of their constitutive human values and cultural meanings.

As if surprised, much conventional research about climate change is now devoted to finding ways of restoring these lost values and meanings (e.g. Adger *et al.*, 2010), to reconnect purified climate indicators with local practices and to re-situate and re-contextualise putative future climates in unique geographical settings (e.g. Wilbanks and Kates, 1999). This explains the attention given in such circles to the arcane practices of climate model 'down-scaling' and to the growing urgency for understanding adaptive practices and resilient coping

---

51 There are other ways of capturing the physical essence of climate – vegetation, ice cover, soil moisture, etc. – yet these are largely derivative; similar reservations would apply to these descriptors.

strategies. It also explains the late, somewhat sheepish discovery by non-anthropological climate change researchers of local indigenous (climatic) knowledge (e.g. Berkes *et al.*, 2000).

There is much geographical work to be done to repair the damage that has been thus caused. Beyond the advocacy of wider cultural readings of climate (as above), this second area of activity asks geographers to play their full part – alongside anthropologists, psychologists, sociologists, literary critics and ecologists – in ensuring local meanings of climate retain their integrity and are more faithfully indexed to the physical dimensions of weather: 'situating [climate change] within a relational context that may include the places people live, their histories, daily lives, cultures or values' (Slocum, 2004: 416). Such research is needed as much, perhaps more, in countries of the North and West as in countries of the South and East. Survey after survey has shown that by constructing climate change as a global problem, one that is distanced and un-situated relative to an individual's mental world, we make it easy for citizens to verbalise superficial concern with the problem, but a concern belied by little enthusiasm for behavioural change (e.g. Slocum, 2004; Lorenzoni *et al.*, 2007). By 'de-culturing' climate – i.e., purifying climate and letting it travel across scales detached from its cultural anchors – we have contributed to conditions that yield psychological dissonance in individuals: the contradictions between what people say about climate change and how they act (Stoll-Kleeman *et al.*, 2001).

Geographers are well placed to do the imaginative yet meticulous work of revealing the local roots of climate meanings and in then finding ways of allowing climate to travel and cross scales without losing these essential anchors and narratives.[52] The study of Nuie (reported in Adger *et al.*, 2010) and the ways in which reactions to climate change and sea-level rise are embedded in much wider stories about the economic, cultural and demographic aspirations of a Pacific island is just one illustration of why universal and globalised constructions of climate will not connect with the local. Jennings' study (2009) of community risk perceptions and climate change discourses in Boscastle in Cornwall, following the floods of 2004, is another. 'The view from no-where' (Shapin, 1998) has no local resonance; climate, and hence climate change, must always be viewed from somewhere. Another example of a geographer at work in this mode is the study by John Thornes (2008) in which he uses visual representations of the atmosphere in the work of Constable, Monet and Eliasson to understand the cultural symbolism of the skies, weather and climate, symbolism which still resonates today in local sub-cultures of Western Europe.

52 This has some similarities to the idea of 'memory mapping' in which writers explore people and places and the relations between them, combining fiction, history, traveller's tales, autobiography, anecdote, aesthetics, antiquarianism, conversation, and memoir (University of Essex, Maria Warner, www.vam.ac.uk/activ_events/adult_resources/memory_maps).

Spatial ordering of climate change knowledge

The third geographical project I propose as urgent is to scrutinise the knowledge claims made by science about climate change, most notably the various assessments of the IPCC. This is part of the bigger project of 'putting science in its place' as argued in David Livingstone's eponymous book. This project that should be undertaken by deploying the insights and tools of the geographer to see how knowledge is spatially produced and consumed and how it travels between sites of production and consumption ... 'in the consumption of science, as in its production, a distinctive regionalism manifests itself' (Livingstone, 2003: 123).

During its 20 year history, the IPCC has been examined critically from a number of different standpoints: dissecting its 1980s origins (Franz, 1997), revealing its norms, practices and self-governance (Agrawala, 1998), debating the role of consensus in its assessments (Demeritt, 2001), and tracing the relationship of its institutional function and knowledge claims to emerging ideas of global environmental governance (Miller, 2004). But other questions remain about the status of climate change knowledge synthesised by the IPCC, questions which emerge from the agendas raised by the new geographers of science (e.g. Powell, 2007). As Sheila Jasanoff has shown in many of her writings (e.g. Jasanoff, 2004; 2005), knowledge that is claimed by its producers to have universal authority is received and interpreted very differently in different political and cultural settings. Revealing the localisation and spatialisation of knowledge thus becomes central for understanding both the acceptance and resistance that is shown towards the knowledge claims of the IPCC.

The IPCC presents its reports to the world as the 'consensus view of the leading climate change experts in the world'. But how are the contours of this knowledge shaped? How localised are the sites of climate change knowledge production and how well does this knowledge travel? This is not just about the very small cabal of government scientists and bureaucrats who constitute the IPCC governing bureau and who determine the structure and content of the various assessments. Nor is this just about the processes and outcomes by which experts are entrained in the assessment activity.

These questions of personnel and their geographical settings should indeed be examined. But there also needs to be attention paid to the peculiarities of the production sites of the primary knowledge assessed: the exclusive network of climate modelling centres that exerts power over descriptions of future climate; the voids on the Earth's surface where few or no observations of climate, phonology or tide variations are made; the digital circuitry and laboratory practices which transform millions of meteorological measurements into (just) a very small number of indices which capture the state of the Earth's climate in a single temperature register. An understanding of the ways in which space modulates these processes of production, and the alternative ways in which this construction process might work, is essential for scrutinising claims of credibility and legitimacy. Some technical self-policing by the IPCC has been undertaken with regard to questions of uncertainty characterisation and peer review (e.g. Edwards and

Schneider, 2001), but geographers of science have not yet embarked on a more penetrating analysis.

Geography plays a central role not only in the production of IPCC knowledge, but also crucially in its consumption. The danger here is that uncritical submission to the globalised knowledge claims of an elitist and labyrinthine institution closes down spaces in which the negotiation of politics might get to work. As pointed out earlier, the deeply materialist register in which climate change is currently marked, sucks out from climate most of its rich and culturally diverse significance. As Cathleen Fogel states, 'the global knowledge that the IPCC produces helps governments erect and then justify their simplified constructions of people and nature, and the institutions based on them' (Fogel, 2004: 109). But beyond this we need to consider how these impoverished constructions of climate change knowledge are variously consumed in different institutional and cultural settings. What counts as authentic public knowledge in one state, might have much weaker traction within another (Jasanoff, 2004). The consensus science of the IPCC might look persuasive from the centralised sites of production. The views from the peripheries of space, of power and of culture – the very places where knowledge is consumed – look very different. We need to understand this story and tell it widely.

Conclusion

This essay has expressed concern that climate change is not making sense to us: we have universalised the idea of climate, detached it from its cultural settings and failed to read the ways in which the knowledge claims emerging from climate science change meaning as they travel. It has argued that geographers have a unique role to play in repairing the damage done – not restoring the physical climate in the literal sense implied by the rhetoric of climate policy, but repairing our idea of what climate means in different places and to different peoples and at different times. By dissolving the strained boundaries between Nature and Culture, by revealing that knowledge and scale are co-dependent, by disclosing the spatial contingencies of climate change knowledge, geographers can offer a different reading of climate change. It will be a reading that is more engaging and human than the one currently on offer.

Why does this matter? It matters because the dominating construction of climate change as an overly physical phenomenon readily allows climate change to be appropriated uncritically in support of an expanding range of ideologies ... of green colonialism (e.g. Agarwal and Narain, 1991), of the commodification of Nature (e.g. Thornes and Randalls, 2007), of natural security (e.g. Sindico, 2007), of celebrity (e.g. Boykoff and Goodman, 2009), and many others. These creeds may or may not be undesirable. But the materialising and globalising properties of the existing framing of climate change – what I call the de-culturing of climate – endow climate change with a near infinite plasticity. Climate change becomes a malleable envoy enlisted in support of too many rulers.

Geographers, as do historians of science (e.g. Latour, 1993) and anthropologists (e.g. Cruikshank, 2005), understand that meaning is lost once the physical and the cultural are cleaved. By purifying climate we de-stabilise the idea of climate, allowing it to be appropriated in support of one-eyed agendas. We cannot allow this to happen. We need first to understand why we disagree about climate change – and we disagree exactly because of the power of the more imaginative perspectives on climate hinted at in this essay – before we can put the idea of climate change more assuredly to work in our local, national and international politics. By stripping climate of its flowing cultural and psychological symbolism, by ignoring the multiple meanings of climate, we are in danger of letting the idea of climate change get out of control.

And why does any of this matter for geography and geographers? For too long geographers have forgotten that climate is an idea that emerges from the heart of the discipline – or rather that geography lies at the heart of the idea of climate.[53] Climate is an idea which encapsulates the immersion of the physical with the cultural, in which local and global dynamics interweave and where the memory of the past meets the possibilities of the future.[54] Embarrassed in the first half of the twentieth century by the naive determinists, geographers became increasingly happy to leave climate well alone. Distancing themselves from the idea, climate was left first to the physical geographers, who in turn handed it over to the meteorologists who most recently have been usurped by the Earth system scientists. This distancing has not been helpful. The time is now right to welcome climate back into the intellectual family of geography. Rather than joining in the project to conquer climate, geographers should be announcing a celebration of climate even as we become active agents in its future. Climate and humanity have travelled a long way together; let us not give up now.

53 The Greek word *klima* means 'slope' or 'incline' referring to the dependence of incident solar radiation on latitude, a geographical concept.
54 Scott Huler's (2004) philosophical biography of the Beaufort wind-scale, in which science and poetry travel through changing cultures, exemplifies this conception of climate.

19 Mapping climate change knowledge

January 2010

Editorial essay written for the launch
issue of the review journal *Wiley
Interdisciplinary Reviews (WIREs):
Climate Change*[55]

New maps of climate change knowledge

The relationship between climate and society has always been dynamic. Physical climates have played important roles in the biological and social evolution of human beings and fashioned the biogeophysical systems which provide the goods and services from which human economic and cultural life emerges. Similarly, humans have imposed themselves upon climate, both in the ways in which climates have been imagined, studied, and articulated in thought and language and in the growing influences that human actions have had on the physical operations of climate.

This relationship between climate and society has been characterised throughout history and pre-history by both creativity and fear (Hulme, 2008). Climates have offered societies productive resources and stimulated new technologies. They have also moulded personal and collective identities and inspired artists and story-tellers. But climates have presented societies with risks and danger too. They have induced fear about survival and presaged anxiety about the future. This intense relationship between climate and society is now even more intimate as the actions on a global scale of a burgeoning humanity are changing in unprecedented ways the physical properties of the climate system, also on a global scale. Future weather will not be like past weather; future climates will not be like past climates.

Climate is not what it was

Human societies have long worried about this possibility and now the knowledge claims of the climate sciences (IPCC, 2007) have given us new reasons to be concerned about the future and a new language through which to express such

55 Hulme, M. (2010) Mapping climate change knowledge – an editorial for *Wiley Interdisciplinary Reviews Climate Change WIREs Climate Change* 1(1), 1–7. This article has been slightly edited to remove journal-related editorial guidance.

concerns. At the same time as the physical processes of climate are being altered, the very idea of climate change is now part of cultural life and is itself changing societies in novel ways. Humanity is firmly embedded within the functioning of the climate system. Yet we have only tentative understanding of the implications of such a new role and only limited means at our disposal to exercise purposeful, as opposed to inadvertent, agency.

These physical manifestations and cultural representations of climate change are interacting in ways that have no historical analogues from which we can learn. And they are doing so in ways that continue to surprise us. The past, through historic emissions of greenhouse gases, is constraining the future in new ways that are still only crudely understood, whereas the future, through scientific predictions and artistic depictions of climates to come, is making new incursions into the present. The idea of climate change is a consequence of this interpenetration of past, present, and future and is acting as a powerful and novel motor for cultural change.

These emergent physical–cultural expressions of climate change present extraordinary challenges to human societies and to the individuals and institutions that comprise them. They are challenges, too, that are reflected in the organisation, practices, and productivity of the knowledge communities. Climate change – although starting as a rather esoteric research question for natural scientists in the middle decades of last century – has now entrained researchers of all varieties and of different instincts. Their research is interacting with politicians, entrepreneurs, celebrities, campaigners, engineers, priests, and citizens in an enlarging search for understanding, for solutions and, ultimately, for security.

Climate change, then, is having to be understood both as physical change – to the planetary systems which create weather, and to biogeophysical environments around the world that are shaped by the weather – and increasingly as an idea that is changing society and the way people think about the future. Researchers have to understand and illuminate the ways these different facets of the phenomenon are shaping each other. Social actions are changing the climate of the future, or at least constraining it, just as physical climates – and simulated virtual climates of the future – are changing society in the present. It is now the turn of climate to reveal the deep entanglement of nature and culture (Yearley, 2006; Hulme, 2009).

The research challenges that arise from this interpenetration of climate and society are made more difficult by the differing research cultures and methods among the disciplines that are contributing to these quests. Witness the recent controversies over the validity of the economic assumptions built into the latest assessments of climate change (Stern, 2007; Garnaut, 2008) the different ways in which climate modelling uncertainties should be characterised and communicated (Dessai *et al.*, 2007), the contrasting positions adopted over the necessity and efficacy of various forms of geo-engineering (Schneider, 2008), or the political adequacy of the Kyoto Protocol (Prins and Rayner, 2007). 'Science' – in a broad interpretation of the practice – does not speak with one voice on climate change.

These examples of contestation emphasise the need for stronger interactions between the traditional climate disciplines of meteorology, oceanography, ecology, and economics on the one hand, and the social sciences and the interpretative humanities on the other. The perspectives and contributions of these latter disciplines are now probably more important than the climate sciences to ongoing public and policy debates. As John Sterman has recently observed: 'There is no purely technical solution to climate change ... for public policy to be grounded in the hard-won results of climate science, we must now turn our attention to the dynamics of social and political change' (Sterman, 2008: 533).

The WIREs concept

It is with the intention of capturing, reflecting, and commentating on this scientific and cultural dynamism that the new concept of Wiley Interdisciplinary Reviews (WIREs) has been applied to climate change. This new journal – *WIREs Climate Change* – is thus launched.

WIREs are a generic new publishing model from Wiley-Blackwell, launched in 2009 and applied to a growing number of inter-disciplinary knowledge domains such as Systems Biology and Medicine, Computational Statistics, and Nanomedicine and Nanobiotechnology. The editorial goal of the WIREs is to emphasise the importance of inter-disciplinarity in science and to support cross-disciplinary collaborative efforts in research and education. The WIREs are not new journals per se; rather, they are hybrid publications that combine the most powerful features of traditional reference works and of review journals in a format designed to exploit the full potential of online publishing. WIREs focus on high-profile, well-funded research areas at the interfaces of the traditional disciplines. They emphasise collaborative and integrative approaches to scientific research, presenting cutting-edge science from a multidisciplinary perspective. They operate as serial publications so that they can benefit from full abstracting and indexing and, especially, impact factors. And they result in a highly structured, comprehensive coverage of a field of knowledge, adopting a common 'templated' editorial format and structure that maximizes quality and consistency within and between the works.

The goal of *WIREs Climate Change* is therefore to facilitate and enhance the introduction and expansion of knowledge about climate change across disciplinary boundaries. It is to promote fruitful new discourses and mutual learning about how climate change is conceptualised, analysed, and communicated in different research traditions. As with the other WIREs projects, *WIREs Climate Change* achieves these goals by performing the functions of a review journal, a dynamic online reference work, and a platform for synthesising and catalysing new inter-disciplinary contributions to our understanding of climate change.

If we examine the various international academic journals that are centred on the idea of climate or climate change, there is no other journal that simultaneously performs these roles. A search through all listed academic journal titles reveals that there are currently 14 journals that have the string 'climat' in their title. This search captures the words 'climate', 'climatic', and 'climatology' and therefore captures the journals whose primary subject matter is the study of climate or climate change. (Of course there are a number of other journals that have traditionally published high-profile work on climate change – *Nature* and *Science* to name but two – but it is instructive in this instance to focus just on those journals whose first and dominant remit is climate).

WIREs Climate Change thus becomes the 15th such journal. Of these 14 'climate journals', 8 have been launched since the establishment of the Intergovernmental Panel on Climate Change (IPCC) in 1988 and 5 of them have been launched since 2005. The disciplinary scope of these climate journals has broadened somewhat in recent years beyond meteorology, climatology, and quaternary science. New journals have been launched recently connecting climate change with development studies (e.g., *Climate and Development*, 2009), with policy and legal sciences (e.g., *Climate Policy*, 2001), and with general social science (e.g., *Weather, Climate, Society*, 2009). Yet with the exception of *Climatic Change* (1977), none of them could claim to be expansively inter-disciplinary.

There are research journals (e.g., *Journal of Climate*, *Climate Policy*) that publish new contributions to knowledge about disciplinary aspects of climate change, and there are some research journals (e.g., *Climatic Change*, *Global Environmental Change*) that do so across disciplinary boundaries. Some of these journals may occasionally publish unsolicited or invited review articles about climate change topics, but there is no broadly based climate journal devoted exclusively to review-type content. And there is no journal that aspires to provide a comprehensive and structured coverage of the full diversity of academic thinking about climate change – from the ontological status of climate to climate system dynamics.

WIREs Climate Change therefore offers something new compared with existing academic climate journals. It also offers an additional service to the knowledge community compared to other major compilations of climate change knowledge such as various climate change assessment reports and print or online encyclopaedias.

The IPCC is one obvious benchmark for comparison. The IPCC has been a unique and pioneering institution for bringing scientific knowledge to bear on an important public policy issue – namely, climate change. Through its various assessment reports prepared at 5 or 6 yearly intervals, the IPCC entrains considerable numbers of experts from around the world in surveying, evaluating, and assessing published knowledge about climate change. As an inter-governmental activity operating under a United Nations mandate, the IPCC is charged with seeking consensus within and across disciplines and also, ultimately, with securing formal agreement between the academy and governments through line-by-line approval of the summary for policymakers. IPCC knowledge is thus co-produced knowledge (cf. Jasanoff, 2004). This is both its strength and its weakness.

WIREs Climate Change offers a different platform for knowledge synthesis and dissemination. The scope and disciplinary reach of this journal is considerably greater than that of the IPCC and, indeed, greater than most other climate change assessments and journals (see above). In designing the intellectual structure of *WIREs Climate Change*, we have de-privileged natural science and economics in telling the story of climate change and introduced stronger contributions from the social sciences and humanities. Thus, *WIREs Climate Change* gives considerable weight to reviewing knowledge about climate change from the perspectives of history, psychology, sociology, ethics, and science and technology studies, areas of knowledge that are not well represented in the IPCC structure (Bjurström and Polk, 2011).

WIREs Climate Change is also able to embrace and reflect disagreement and contestation in the understanding of climate change without the constraint of needing to work towards consensus. The journal adopts conventional practices of academic peer-review in ensuring quality in the material published but, unlike the IPCC, this peer-review does not extend to governmental and extra-governmental interests, nor does it require a process of line-by-line approval of key summary texts and messages. As highlighted above, it is naive to think that 'science' in its various disciplinary manifestations speaks with one voice about climate change, and it is important for scholars and practitioners to be given access to the origins and nature of disputed or plural knowledge. If climate change is an exemplar of what Silvio Funtowicz and Jerry Ravetz have called 'post-normal science' (Funtowicz and Ravetz, 1993), then revealing the origins and reasons for disputed and uncertain knowledge is as important for public policy as is constructing a consensus of 'agreed' knowledge.

If *WIREs Climate Change* is uniquely positioned in relation to conventional research journals and to knowledge assessments such as the IPCC, how does it compare with another – more traditional – mode of knowledge mapping: the encyclopaedia? The leading print encyclopaedia covering climate change is probably the *Oxford Encyclopedia of Climate and Weather* (Schneider, 1996), with a second edition forthcoming (Schneider, 2011). Another, more recent, contribution is Sage's *Encyclopedia of Global Warming and Climate Change* (Philander, 2008) and many aspects of climate change are also covered in other sectoral encyclopaedias such as the *Encyclopedia of Global Change* (Goudie and Cuff, 2001) or the *Encyclopedia of Paleoclimatology and Ancient Environments* (Gornitz, 2008). There are also a growing number of online encyclopaedias, from the generic open-source Wikipedia (en.wikipedia.org) to the new *Encyclopedia of Earth* (www.eoearth.org) from the Environmental Information Coalition. Then there is also the poorer cousin of the encyclopaedia – the dictionary – for example, the *Dictionary of Physical Geography* (Thomas and Goudie, 2000) or the *Dictionary of Environmental Governance* (Saunier and Meganck, 2007).

WIREs Climate Change advances the traditional role of encyclopaedias – and the dictionary – in two ways. It offers a dynamic, serialised platform for the systematic presentation of knowledge about climate change in contrast to the frozen-in-time depictions of knowledge offered by traditional print encyclopaedias. By building incrementally a consolidated body of knowledge, frequently refreshed and expanded through newly written content, *WIREs Climate Change* offers more than the print encyclopaedia can offer and all that is offered by the online encyclopaedias. But it also offers more. It achieves its goal through inviting leading researchers to write review-type articles that tackle important themes, methods, and emergent topics, rather than through one-off entries describing or defining static concepts. *WIREs Climate Change* therefore brings the immediacy and peer-reviewed authority of the research journal medium alongside the flexibility of online reference works.

Structuring *WIREs Climate Change*

How have we applied this generic WIREs publishing model to knowledge about climate change? We can summarise our approach by commenting on two aspects of our structuring: the arrangement of content and the goal of inter-disciplinarity.

The content of *WIREs Climate Change* is organised around 14 knowledge domains. In structuring the content in this way we have steered a path between, on the one hand, merely adopting well-established categories – such as 'observations', 'climate modelling', and 'adaptation' as used in the IPCC assessments – and, on the other, creating a completely new set of categories. In this we have followed in the path pursued by the New Encyclopaedia Project in seeking to tension the mapping of knowledge around 'creative and innovative dimensions of research ... [with] ...the necessity of teaching and writing in succinct and accessible ways' (Featherstone and Venn, 2006: 16).

We have balanced the predominance of the natural sciences and economics in many previous assessments of climate change with knowledge domains more strongly rooted in history, geography, psychology, ethics, and sociology. This structure gives greater salience to the development of inter-disciplinary knowledge about climate change (e.g., the domain of 'Integrated assessment of climate change') and also to areas of fruitful emerging research rarely seen in IPCC reports (e.g., 'Climate, nature and ethics' and 'Perceptions, behaviour and communication of climate change'). The category 'Climate, history, society and culture' is included to allow historical understanding of the changing relationships between climate and society to set the context for subsequent research about what climate change means for the future. And the category 'The social status of climate change knowledge' recognises that one of the central questions affecting how climate change is debated in public concerns the status – the legitimacy, credibility, and saliency – of knowledge claims about climate change. We need to understand how such knowledge comes into being and what types of authority it carries when it circulates through society.

Inter-disciplinary research is a widely lauded aspiration and a frequently claimed necessity in order to advance human understanding of a wide range of troubling, exciting, or otherwise important phenomena. Climate change is without doubt one of these, as was recognised as far back as 1977 by Stephen Schneider in his opening editorial of the journal *Climatic Change* (Schneider, 1977). Yet it is far easier to use 'inter-disciplinary' in a rhetorical or descriptive sense than it is to define what constitutes inter-disciplinary research in theory and, even more so, than it is to achieve it in practice. There is much inertia both in the practices of science and in the conduct of the academy to achieving the nirvana of inter-disciplinarity (see Lowe and Phillipson (2006) and Robinson (2008) for some helpful reflections on this).

WIREs Climate Change approaches this elusive property of inter-disciplinarity at three different levels: at a meta-level for the journal as a whole; at a domain level for different thematic areas of knowledge; and at the level of individual articles. For the journal as a whole, *WIREs Climate Change* encompasses knowledge about climate change drawn from a very broad range of disciplines. Organising this knowledge systematically across a single publishing project so that relationships and conflicts between different areas of knowledge are made visible is a form of inter-disciplinarity, if only the weakest form; indeed, this should more properly be described as facilitating simply a multidisciplinary approach to mapping knowledge.

At the structural level, each of the 14 domains designated in *WIREs Climate Change* has a potential fruitfulness through allowing inter-disciplinary reviews to flourish. Each domain engages with a number of different disciplines – for example, the domain 'Perception, behaviour and communication of climate change' draws upon psychology, sociology, risk and media studies – and articles in each domain can draw upon different disciplinary methodologies in composing their reviews. Individual review articles can therefore be solicited which engage with the synergies and conflicts that emerge at the junctures of different disciplinary theorising and practice. This too is another facet of inter-disciplinarity. There will be occasions when this inter-disciplinary goal is further enhanced through co-authorship of articles, where co-authors are writing from within different knowledge traditions and yet where they seek to offer reconciliation between them.

The above illustrations focus on the practice of inter-disciplinarity in the writing of research reviews. *WIREs Climate Change* also affords opportunity for the advancement of inter-disciplinarity through the reading of review articles. By mapping climate change knowledge across such heterogeneous terrain and through offering navigable online access to readers of this kaleidoscopic map, the journal will introduce students, researchers, and practitioners to ways of conceiving, analysing, and reflecting on climate change with which they may be unfamiliar. If this offer is accepted, it may be a good starting place for readers to embark on a different journey of discovery. An inter-disciplinary understanding of climate change can be cultivated in the mind of the reader as much as it can be constructed through the productive craft of the researcher and author.

Climate change is altering the world around us. It is changing the physical dynamics and configuration of the planet and the distribution and properties of material resources. Maps of world climate will look very different to our grandchildren than they looked to our grandparents. Climate change is also changing the world around us in other ways. Business, political, and social worlds are changing in response to climate change, and for many people climate change is altering their imaginary worlds, the ways in which the future impinges on the present.

WIREs Climate Change is a new journal seeking to reflect these new worlds in the making, and to interact with them, by bringing together into one systematic and dynamic structure emerging knowledge of climate change from across the academic disciplines. It is a new publishing model in that we seek to blend the systematics of a reference work with the dynamics of a research journal, and to do so using the versatility of the online medium for connecting and navigating through complex inter-disciplinary maps of knowledge. We seek to engage readers wishing to access reviews of established knowledge of climate change drawn from disciplinary traditions other than their own. And we seek to stimulate readers with reviews of emerging inter-disciplinary ventures and of debates where the socially engaged nature of climate change research rubs uncomfortably up against knowledge created in the quieter spaces of the laboratory or the study.

20 Meet the humanities

June 2011

An essay published in
Nature Climate Change[56]

An introduction needs to be made between the rich cultural knowledge of social studies and the natural sciences, says Mike Hulme

The Editorial in the first issue of *Nature Climate Change* remarked that 'climate change is now as much a societal problem as a physical one' (Anon., 2011: 1). Although climate is unarguably changing society, social practices are also impacting on the climate. Nature and culture are deeply entangled, and researchers must examine how each is shaping the other. But they are largely failing to do so.

A recent study (Bjurström and Polk, 2011) analysed the disciplinary source literatures of the three working groups of the third assessment report of the Intergovernmental Panel on Climate Change (IPCC). It shows that the cited literature was heavily dominated by natural science disciplines, especially the Earth sciences, while the minority social science content was heavily dominated by economics. Literature from the humanities was virtually absent. The view of climate change thus constructed by the IPCC – and the view which therefore has circulated through societies and influenced policy – is heavily one-sided. Although there may have been some modest broadening of disciplines sourced in the IPCC's fourth assessment report and the forthcoming fifth assessment, the analysis of anthropogenic climate change continues to be dominated by positivist disciplines at the expense of interpretative ones (O'Neill *et al.*, 2010; Nisbet *et al.*, 2010).

This partiality matters profoundly, because such assessments determine the framing of what exactly is the climate change 'problem' that needs to be 'solved', and they set the tone for the human imaginative engagement with climate change. Over its 23-year history, the IPCC has been presented as the authoritative voice of climate science and the global knowledge community. How the idea of climate change is framed by the IPCC therefore carries enormous significance for

56 Hulme, M. (2011) Meet the humanities *Nature Climate Change* 1(4), 177–179.

the subsequent direction, tone and outcome of policy and public debates. As a result of the IPCC's heavy lean on natural sciences and economics, the dominant tropes in climate policy discussions have become 'improving climate predictions' and 'creating new economic policy instruments'; not 'learning from the myths of indigenous cultures' or 're-thinking the value of consumption'.

Steps need to be taken to right the balance. The role of story-telling needs elevating alongside that of fact-finding. The 'two cultures' divide between the arts and sciences in education needs reconciling and collective assessments of climate change knowledge need re-designing.

The humanities are producing journal papers, and indeed special issues, devoted to the topic of climate change (see Box 1). These fields yield knowledge of how people perceive and act on risk based on their cultural and social backgrounds – whether they are risk takers or feel vulnerable to natural disasters, for example (Slovic, 2010). The humanities shed light on the implications of how climate science is represented in the media (Boykoff, 2011; Doyle, 2011). They create an appreciation of the many non-scientific ways in which people sense and interpret the weather (Crate and Nuttall, 2009): indigenous myths and stories can hold valuable information about the timing of seasonal changes; local populations hold a wealth of information about day-of day-to-day, seasonable or decadal weather variability; our memories of long-term weather patterns, even if they don't accurately reflect reality, can affect how we react to scientific accounts of climate change. Furthermore, the humanities can reveal how and why people engage or disengage with different representations of climate change in art,

Box 1 Selected humanities disciplines and their engagement with climate change as illustrated through recent journal special issues on climate change

Discipline	Special issues on climate change
Anthropology	*Anthropology News* 48(9), 2007
Communication studies	*Science Communication* 30(3), 2009
	Environmental Communication 3(2), 2009
Ethics	*Environmental Justice* 2(4), 2009
Historical geography	*Journal of Historical Geography* 35(2), 2009
History of science	*Osiris* 26(1), 2011
Literary criticism	*Oxford Literary Review* 32(1), 2010
Museum Studies	*Museum and Society* 9(2), 2011
Philosophy	*Journal of Social Philosophy* 30(2), 2009
	The Monist 94(3), 2011
Psychology	*American Psychologist* 66(4), 2011
Religious Studies	*Journal for the Study of Religion, Nature and Culture* 6(3), 2012
Sociology	*Theory, Culture and Society* 27(2/3), 2010
	The Sociological Quarterly 52(2), 2011

fiction or performance (Trexler and Johns-Putra, 2011; Yusoff and Gabrys, 2011). The current debate on geo-engineering, for example, has long been explored in science fiction, reanimates ancient Promethean myths about human mastery, and challenges new contemporary myths about 'Mother Earth'.

All these types of knowledge need to be embraced. As communications professor Matthew Nisbet of the American University and colleagues recently argued, we need 'to bridge the great wellsprings of human understanding – including the natural and social sciences, philosophy, religion and the creative arts – to "re-imagine" how we live on Earth' (Nisbet *et al.*, 2010: 330).

The opening-up of climate change to scrutiny from interpretative disciplinary traditions should not be achieved on the terms offered by natural scientists alone. In 2008, the former UK Government Chief Scientific Advisor, Sir David King, accused academics in the humanities of 'shirking the climate change fight' and criticised 'academics in the arts, humanities and social sciences for staying in their disciplinary "comfort zones" and failing to engage with scientists on the problem of climate change' (Corbyn, 2008). Yet this engagement must work both ways. It needs to be acknowledged that the role of arts and humanities is not simply to translate scientific knowledge into public meaning, as though science is the only source of primary knowledge. Neither will the humanities 'solve' the problem of climate change. Humanities research is accumulative rather than progressive. In the words of professor of philosophy Nicholas Davey from the University of Dundee, it 'thickens and extends an understanding of the issues involved' (Davey, 2011: 305).

So what is to change, and how? First, the importance of story-making and story-telling around climate change needs elevating alongside that of fact-finding. Stories are the way that humans make sense of change, and the humanities understand the practices of story-telling very well. At the Open University, for example, the Creative Climate project[57] is a new online venture for collecting global diary entries about how climate is affecting individuals, groups, and even streets or insects. Such anthropological efforts need to sit alongside satellite monitoring and precipitation databases, as providing equally essential data in the search for understanding.

Second, we need to pursue a higher-education agenda in which the apparent distance between the arts and sciences is narrowed, if not eliminated. This will not suit all temperaments. The contemporary political orthodoxy is that investment in science, technology, engineering and maths (the STEM disciplines) provides the most assured basis for securing future economic vibrancy, social well-being and environmental protection. Yet the STEM disciplines by themselves carry a hubris which they seemingly cannot shake off. On their own they are inadequate for tackling 'wicked' problems such as climate change (Brown *et al.*, 2010). Reconciliation between the two demands an educational ethos sympathetic

---

57 http://www.open.ac.uk/openlearn/nature-environment/the-environment/creative-climate

to the idea that humans live simultaneously in both material and imaginative worlds. The new MA/MSc in Environmental Sciences and Humanities being developed at the University of East Anglia is such an example.

Finally, knowledge assessments as exemplified by the IPCC need a fundamental re-think. Crafting increasingly consensual reports of scientific knowledge, or levering more engineering and technology, will alone never open up pathways from research to the public imagination or the execution of policy. It is not too soon to start a conversation about whether the world in 2020 really needs a sixth assessment report of the IPCC that merely echoes the previous five, or whether it needs something fundamentally different. Although it is valuable to have a single, global-level, government-owned assessment of how climate is changing, for designing adaptation and mitigation interventions it would be more valuable to have regional-level, non-governmental assessments with locally developed protocols that are sensitive to cultural conditions.

A better balance is achievable. This October, countries in the Convention on Biological Diversity plan to launch a proposed Intergovernmental Platform on Biodiversity and Ecosystem Services (IPBES) – a body partly inspired by the IPCC, but designed from the outset to allow for the respect and hearing of insights from the humanities and non-scientific knowledge. The *Busan Outcome* for IPBES says the organisation should 'recognize and respect contribution of indigenous and local knowledge', 'take an inter- and multi-disciplinary approach that incorporates all relevant disciplines including social and natural sciences' and generate 'new knowledge by dialogues with stakeholders' (UNEP, 2010).

The positivist disciplines are ill-suited to engaging with and articulating the deeper human search for values, purpose and meaning – and yet this search is exactly where humanity's new entanglement with global climate is taking us. To shed new light on the multiple meanings of climate change in diverse human cultures, and to create new entry points for policy innovation, the interpretative social sciences, arts and humanities need new spaces for meeting as equals with the positivist sciences.

Section four

Culture

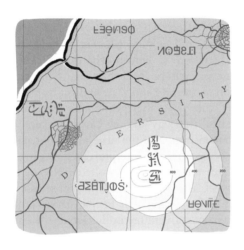

Introduction

In this section I bring together six articles which in various ways reveal the entanglement of climate and culture, whether through cultural filters which condition our interpretation of climate change or else through the inter-dependency of climate and culture. In the first essay (Chapter 21), written back in 1990 for the Christian magazine *Third Way*, I reflect on the implications of what was then commonly known as 'the greenhouse effect' using two lenses: my interpretation of biblical theology and my Christian faith. The significance of this piece is in its demonstration of the need for extra-scientific reasoning – in this case a Christian perspective – to interpret climate change in personal, ethical and cultural terms. This was a position I was to develop much more explicitly in my later work, by which time the idea of the post-secular society and the re-emergence of religion in European public life had gained ground.

A different engagement with culture is illustrated in the second item included here (Chapter 22), a short essay inspired by John Constable's 1822 'cloud study' and invited by the London-based arts magazine *Tate Etc*. I used clouds as a metaphor for the intractability of climate change – complex, constantly shifting and beyond deterministic prediction. Written in early 2007, this essay offered some signposts towards the ideas I would develop in my book *Why We Disagree About Climate Change*.

The next two items (Chapters 23 and 24) both challenge the claim that climate is a powerful, even primary, determinant of human character and culture. Taken to extremes this claim leads to the excesses of climatic determinism which has a long and, at times, unsavoury history. The prompt for the essay written in 2007 for the on-line *OpenDemocracy* forum was my concern that climatic determinism was making a comeback under the guise of 'climate change and security'. This followed a spate of reports which seemed to claim that climate change was a cause of war and geopolitical insecurity and also followed the UN Security Council debate on climate change prompted by the British Government in May 2007.

The spectre of climate determinism and its ideological history was also in my mind as I reviewed in 2010 Wolfgang Behringer's book *A Cultural History of Climate* for the on-line journal *Reviews in History* (Chapter 24). I used Behringer's book to explain how difficult a task the historian has in discerning the agency of climate in human affairs, seeking to avoid the two extremes of reducing history to climate on the one hand or else of ignoring climate completely on the other. By extension, this difficulty is just as great when it comes to speculations about the future impact of climate change on human culture, an argument I develop at length in an article written the following year for the history of science journal *Osiris* (Hulme, 2011).

One way to avoid these two extreme heuristic positions is explored in my next selection (Chapter 25), a short article written for the German-published journal *Nature and Culture*. I argue here that the now inescapable human influence on climate should make us re-think our idea of climate as some physical boundary condition for human action. Rather, since we are now continually re-shaping our climate and weather – even if inadvertently – we should embrace such novelty and learn to live with the consequences, just as we do in relation to our co-created ecosystems. The boundaries between the natural and the artifactual are breaking down in relation to climate, just as they have already done so in relation to ecosystems, the human body and chemical matter.

The final selection for this section also engages with the idea of culture, this time through the lens of Kari Marie Norgaard's book *Living in Denial*. My review of her book appeared in *Nature Climate Change* in 2011 and I praised the way in which she explored the idea of 'climate change denial' through an ethnographic study. How and why different people embrace or distance themselves from the idea of human-induced climate change is a matter for understanding, not for polemical denigration of one's opponents. Through studying culture and climate in a small Norwegian town, Norgaard illuminates the reasons why cultural beliefs and practices are intimately bound up in the ways in which people make sense of change in the world around them, whether this be climatic, environmental or any other sort of change.

21 Rainbows in the greenhouse

May 1990

Essay written for the Christian magazine, *Third Way*[58]

> Then God said to Noah and to his sons, 'Behold, I establish my covenant with you and your descendants after you, and with you every living creature that is with you … never again shall there be a flood to destroy the earth.
>
> (Genesis 9.8–11)

Ninety per cent of non-saline water on our planet is stored as ice. Some of this is held in mountain glaciers, some sea ice and icebergs, but the vast majority is held in the two great ice caps of Antarctica and Greenland. Should all of this ice melt, the average global sea-level would be between 50 and 60 metres higher than it is now – not a flood to destroy the Earth but enough to affect the lives of 80 per cent of the world's population. This is impossible within anything less than several thousand years. More plausible, although still unlikely, is that by the end of the next century (when our grandchildren are retiring) the substantial melting of one part of Antarctica will lead to a rise in average global sea-level of between four and five metres. This would submerge the Nile Delta, inundate stretches of the eastern American coastline, and eradicate some island nations in the Indian and Pacific Oceans. The most likely assessment, however, of sea level change over the next 100 years is that it could rise between 40 and 60cm – certainly not a flood to destroy the Earth, but sufficient to pose intractable problems for island nations and low-lying coastal states.

Industrial gases

The cause of this potential future sea-level rise is global warming, or what is now ubiquitously termed the greenhouse effect. Greenhouse gases, primarily carbon monoxide, methane, nitrous oxide, chlorofluorocarbons, and low-level ozone, are now generated to an unprecedented volume by an expanding and increasingly industrialised humanity. It is this inexorable rise in the atmospheric concentration

58 Hulme, M. (1990) Rainbows in the greenhouse *Third Way*, May, 22–24.

of the gases, perhaps unfortunately all colourless, odourless and tasteless, which drives global warming.

The idea of the greenhouse effect has become firmly established. Advertising for products as varied as cars, shampoo, and electricity refer to it freely, it is one of the touchstones by which policy in Whitehall is now assessed, and it is the subject of much popular speculation concerning Britain's future climate.

So, as 'descendants of Noah' we live under God's covenant with the living Earth, and as future citizens of the twenty-first century we live with the prospect of the greenhouse effect. Is the Church, God's people on Earth, bothered? Chris Patten, Secretary of State for the Environment, recently interviewed on the subject of environmental policy, stated 'what is at stake is the future of our planet, which is clearly a moral issue'.[59] If this is true then the Church can no more turn an uninterested eye away from the greenhouse effect and its causes than it can to the moral issues of homelessness, famine, or AIDS. But there has been astonishingly little Christian comment on it. If, as recently stated by the Worldwatch Institute in Washington DC,[60] securing our climate future is now on a par with consolidating superpower security as the two leading global issues facing humanity, why are Christians receiving so little input on the problem from the Church? We need to examine exactly what we do and do not know about the greenhouse effect, how it challenges us, and the range of response open to us.

What we do know

We depend on the greenhouse effect for our survival on this planet. If our atmosphere contained no greenhouse gases, merely oxygen, nitrogen and trace volumes of inert gases, the average global temperature would be 30 degrees colder than it is today and humanity would die, for all the energy the Earth receives from the sun is ultimately re-radiated to space. The exact temperature of the planet is therefore determined by how long this energy remains within the atmosphere. The longer it stays, the warmer the Earth. Greenhouse gases, although of low concentration in the atmosphere, delay the return of energy to space, thus raising the surface temperature. Loading the atmosphere with more greenhouse gases is like turning up the thermostat on a storage heater.

What has been occurring over the last 100 to 200 years is an enhancement of the natural planetary greenhouse effect by an increase in the atmospheric concentrations of greenhouse gases. We are causing this by a variety of social and economic activities which are unique to our industrial era. The single most significant of these activities is the combustion of fossil fuel for electricity generation and for vehicle mobility which releases, among other gases, carbon dioxide.

59 *On the record* BBC1, November 26 1989.
60 Worldwatch Institute (1989) *State of the world 1989* Worldwatch Institute, Washington DC.

This rise is global in scale. Irrespective of the emission of gases being in the past concentrated toward the mid-latitudes of the northern hemisphere, greenhouse gases mix rapidly throughout the atmosphere. The contribution of different nations to the global rise, however, is highly uneven. An average American or Pole injects more than four tons of carbon into the atmosphere per year, a Briton just over three tons, and a Kenyan about one ton.[61] To a greater or lesser extent we all contribute to the greenhouse effect and our demand for car transport, manufactured goods, and cheap energy all adds to the prevailing level of greenhouse gases in the global atmosphere. A capitalist petrochemical plant in the UK is as deadly as a burning rainforest in Brazil, or a communist coalmine in China. This is the ultimate in collective responsibility. Our lifestyles are atmosphere-polluting and climate-altering.

What we don't know

Since science seems clear on the question of rising greenhouse gas concentrations and their direct effect on air temperature, then historically we should already have witnessed a rise in the global-mean temperature. By compiling thousands of individual temperature readings from around the world we can estimate the average temperature of our planet to within about 0.1°C. Such readings extend back to the mid-nineteenth century and when analysed suggest an overall rise of about 0.5°C over the last 100 years. This warming has not been even either through time or by region. The magnitude of this global warming since the mid-nineteenth century is in line with greenhouse theory, but owing to the natural variability of global-mean temperature it cannot yet be unequivocally attributed to the greenhouse effect. Other intervening mechanisms, such as frequency of volcanic eruptions or the sunspot cycle, could still conceivably account for the observed trends.

The harder question for science to answer concerns the magnitude and speed of future warming, and harder still, the consequences of such warming for global sea-level and other important climate variables such as rainfall, wind speeds, and hurricane frequency. However, there exists a clear consensus among all the independent modelling groups who have addressed the problem that further warming of between 1.5 and 4.5°C is likely by between 2050 and 2100 with a net global increase in rainfall.

Even if all greenhouse gas emissions were stopped tomorrow, the projections show that the planet is committed to a warming of between 0.5 and 1.5°C. Using our analogy of a storage heater this arises because of the time lag between setting the thermostat at a given temperature and the eventual stabilising of the heater at that temperature. The oceans, the Earth's most effective energy storage medium, have not yet reached equilibrium with the current atmosphere

---

61 These figures refer only to the carbon we are unlocking from semi-permanent stores such as coal, oil and gas.

greenhouse gas concentration; such equilibrium will take between another 10 and 50 years to occur. It seems our planet is firmly committed to a substantial warmer climate in the twenty-first century than any century the human race has lived through before.

Not healthy

One question that does *not* arise is that of defacing God's creation. Unlike the destruction of living species, the greenhouse effect does not deplete the richness of God's creation. The full range of climates from arctic winters to topical monsoons will continue to exist in a warmer world. Neither is the greenhouse effect inherently unhealthy. Unlike the related problems of acid rain, urban smog and the depletion of the high-level ozone over the poles, it does not pose any *direct* threat to human health. One may indeed argue that less severe winters in mid-latitudes will save lives, the expansion of agriculture into the tundra regions of Canada and the USSR will benefit food production, warmer nights in the subtopics will reduce the demand for woodfuel and hence forest cutting in these countries.

It is at this point that the complexities of environmental cost/benefit analysis enter our assessment. Such analyses are now applied, commonly if not routinely, to local developments such as a new supermarket, or nationally as in the Channel Tunnel. Rarely, however, have successful cost/benefit analyses of environmental changes been applied internationally, and the greenhouse effect operates on a spatial scale and order of magnitude beyond this. A global-scale cost/benefit analysis is demanded. Some countries will become winners and some losers as it begins to bite. As ever the poorest people and nations will be the most vulnerable. A sea-level change rise will have far more devastating effect on the poor of Bangladesh than on the rich of the Netherlands.

Citizens of the world

This is therefore the ultimate question posed by the greenhouse effect: do we have the capability to manage our planet on a scale never required before, and to adapt to changes on a scale never seen before? The critical issues of economic interdependence raised by the Brandt and Brundtland reports in the early and late 1980s or the crisis in African food production in 1984 and 1985 have also posed serious questions which should be the concern of all the world's citizens. A changing global climate, however, places a hitherto unrealized degree of self-interest upon what we have been told many times before. We are citizens together of the same planet, dependent upon each other for our well-being. 'Who is our neighbour?' the disciples asked Jesus in Matthew, chapter 13. Our neighbour is the person next door. Our neighbour is also the Kenyan farmer who faces crop failure because of changes in seasonal rainfall, the Maldivian fisherman who sees his home swamped by a raising ocean, and the Egyptian cultivator whose delta faces inundation. It is their livelihoods if not their lives that are threatened by our collective actions and in a way more direct than through the impenetrable and rather remote structures of international trade.

If the greenhouse effect creates a new urgency for accepting our global interdependency, how do we respond? There are four possibilities. The first is for both the optimist and the fatalist and requires us to do nothing. Optimists state that humanity has a long history, has lived through at least one ice age, and that our inventiveness and adaptability is our best defence against global warming. This will see us through. Fatalists, likewise, would have us do nothing for they know the planet is doomed whether through nuclear holocaust, a global epidemic, or global warming. A religious expression of such a view is not uncommon: a recent example appeared in the magazine of the Jehovah Witnesses:[62]

> Yet which is more naive ... to hope that mankind will reserve its sad history and take farsighted action to avert (climatic) disaster in the next century, or to believe that God will intervene before it is too late? The Creator has promised in his Word to bring ruin to those ruining the earth (Revelation 11.18). There is ample historical and scientific evidence that he intends to do just that.

The second response would develop a variety of international insurance arrangements. We cannot stop global warming and we know it will generate detriments unevenly among and within nations, but we can ensure that the winners compensate the losers in either finance or kind. The term 'national insurance' would take on a different meaning, especially for nations like the Maldive Islands whose island home would disappear following a 1.5 m rise in sea-level (but can a nation be in any way compensated for its destruction?)

Third is the approach of active adaptation. If global warming is imminent and since such warming will change our physical environment in rather uncertain ways, we should, as deliberate policy, establish greater flexibility within our agricultural resource management, and transport systems. For examples, monocropping is a high-risk strategy in a changing climate and should be replaced by mixed cropping: a truly national watergrid with the capacity for shifting large quantities of water from region to region would provide protection in case of drought: flood barriers such as the Thames Barrage should be capable of protecting not just from a normal storm surge, but from a global rise in sea-level. Such actions are undoubtedly expensive and there would be many who would remain sceptical that such large expenditure was warranted in the face of an uncertain threat. (But one might speculate about the sceptics Noah had to face).

62 *Awake* magazine, September 1989.

A preventive strategy

All the above lines of thought assume that the course upon which we have embarked will not be stopped. The only response which approaches the heart of the problem is one which tackles the emissions of greenhouse gases – the preventive strategy. Such an approach would advocate some obvious and some not so obvious, some relatively straightforward and some intensely complex strategies to pursue. These include large-scale energy conservation, encouragement of collective transport, sustainable forestry in both low and high latitudes, and the stimulation of renewable energy technologies.

One strong reason to favour this preventive strategy is that many of the advocated policies are justifiable in their own right. Should the climate system subsequently surprise us, should our science be shown to be poor, should we discover much greater human inventiveness and adaptability than we currently imagine, then we will have pursued strategies which are beneficial in themselves. Thus energy conservation is economically prudent, collective transport is safer than private transport, sustainable forestry improves the quality of the local environment and renewable energies generate far fewer (if any) non-greenhouse pollutant gases such as sulfur dioxide or nitrogen oxides which are both also implicated in acid rain and photochemical smog.

The greenhouse effect, uncertain in its consequences as it is, is reminding us of how unhealthily we live, and focusing our minds upon alternative lifestyles. Over 10 years ago [Ronald Sider's book] *Rich Christians in an Age of Hunger*[63] was advocating that Christians reappraise their lifestyles because of the global hunger issue. The greenhouse effect is now telling us something very similar in the 1990s. It provides, however, a stronger motive for change; human nature being what it is means that substantial disruption of one's immediate environment appeals more to self-interest than does hunger in distant lands.

Finally, how should the Church address itself to these challenges? The 'end times' syndrome has been alluded to already and such thinking is by no means confined to the Jehovah Witnesses. One will find such viewpoints in most Christian denominations. This view must be firmly dismissed. It is never for us to prejudge God's intervention in history and we must act and plan for tomorrow.

The second problem is a bigger one. The Christian Gospel, perhaps especially in the West, has been highly individualistic in emphasis, both in its assessment of sin and of repentance. This is in contrast to much Old Testament history and theology. The disturbing way we are treating this planet calls not for a collective repentance of individual sin, but for an individual repentance of collective sin. The greenhouse effect is a social problem which would not be solved by a 100 per cent conversion to Christianity. It *would* be solved by a 100 per cent conversion to a climate-friendly lifestyle. The Church seems to have little to say on this.

---

63 Sider, R. (1978) *Rich Christians in an age of hunger* Hodder & Stoughton, London.

The greenhouse effect is an immensely complex issue involving scientific, diplomatic, political and lifestyle perspectives. No wonder any one Christian is slow to comment or indeed to pass judgement on profligate and consumptive lifestyles. But just as the greenhouse effect is now employing the best minds in science, Government and international diplomacy, so too it must occupy the collective mind of the Church.

God's promise that there shall never again be a flood to destroy the Earth is no ground for Christian complacency. Rather, it is our motivation to work actively with God's special planet to ensure that his promise is fulfilled. Good planets may be hard to find, but God died for this one.

22 On John Constable's *Cloud Study* (1822)

Spring 2007

An essay written for *Tate Etc.*, the arts magazine of Tate Britain[64]

To John Constable clouds were the essential frame for his studies of landscape, the 'key note' and the 'standard of scale' against which his compositions were structured. Yet for meteorologists clouds are the most ephemeral of weather phenomena, the hardest to measure objectively and the most difficult to understand in terms of processes of formation and decay. The classification of clouds, following the pioneering nineteenth century work of Luke Howard, is today still as much an art as it is a science. Clouds also have a fascination for humans, as evidenced by the current best-seller *The Cloudspotter's Guide* which provides a grand worldwide survey of the history, science and culture of clouds.

Whilst clouds played a crucial artistic role for John Constable, for scientists trying to understand the way the world's climate system works clouds are the most frustrating and intractable of elements to tackle. They best embody the tentativeness of our predictions of future climate. For example, one of the most powerful computers in the world – the Japanese Earth Simulator, capable of 35 TeraFlops (or 35 trillion individual calculations per second) – is still unable to simulate the growth and decay of individual clouds, reverting to statistical approximations of 'typical cloudiness' over parcels of sky several square kilometres in size. And the full resources of the CERN Laboratory near Geneva – the particle physicist's holy city – is to be deployed next year in a series of experiments to discover whether cosmic rays from the sun can influence cloud formation, and hence help regulate world temperature.

Clouds then remain one of the most persistent sources of the uncertainty that afflicts scientists' predictions of future climate change, an uncertainty often cited by conservative voices in our society that nothing need be done about climate change. Yet climate change is now far more than a discovery of the natural sciences and can no longer be defined, debated and defused through advances in scientific knowledge. Climate change is (as are clouds) as much a cultural phenomenon as a physical reality.

---

64 Hulme, M. (2007) Mike Hulme on John Constable's *Cloud Study* (1822) *Tate Etc.*, Issue 9 (Spring), p.109. © TATE ETC. magazine.

Debates about climate change still defer to the authority of the meteorologists and the Earth system modellers who argue that this tipping point or that climate impact will provide the final piece of evidence to ensure a breakthrough in the international diplomatic negotiations about climate change. Instead, underneath the surface, these negotiations reveal the full complexity, inequality and intractability of a troubled world, a world where different ideologies, cultures, faiths, and economics battle for ascendancy and power.

We disagree about climate change and what to do about it not because the science is still emergent, uncertain or incomplete. It is not because of the ephemeral and intractable nature of clouds. We disagree because everyone has a stake in the problems raised by climate change, and in the solutions offered. To reconcile these different and often competing stakes we need to harness the full array of human sciences, artistic endeavours and civic and political pursuits to make progress – burdening science alone with the responsibility will not make for good science, nor for good policy.

Climate change is not a problem waiting for a solution. Engineers are very useful people, but they are not going to give us the answer here. And climate science will never deliver the certainty about future change nor unambiguously define the probabilities of climate-related risks which will provide the world with the necessary tool-kit to decide what to do. We need a far richer array of intellectual traditions and methods to help analyse and understand the problem of climate change – behavioural psychologists, sociologists, faith leaders, technology analysts, artists, political scientists, to name a few.

And we ultimately must recognise that climate change is the most deeply geo-political, not simply environmental, issue faced by humanity. Climate change will not be 'solved' by science. Yet climate change *can* be reconciled with our human and social evolution, with our endurance on this planet, if we allow it to escape from its scientific ghetto … in this sense climate change can be defused from being a looming threat to mankind and instead can be used as a creative stimulus for a more sustainable pattern of life.

The emotive power of clouds, our fascination with them, and their scientific intractability, make them a good metaphor for the deep complexity of the climate change conundrum. Yet without clouds there is no rain, and without rain, we die.

23 Climate security: the new determinism

20 December 2007

An essay written for *openDemocracy* on-line[65]

There is a new form of climatic determinism on the rise and the allure of this thinking for the naive or for the mischievous is dangerous. It finds its expression in some of the balder claims made about the future impacts of climate change: 180 million people in Africa to die from hunger; 40 per cent of known species to be wiped out; 20 per cent of global GDP to be lost. But such determinism is perhaps at its most insidious when found in the new discourse about climate (in) security. Here are only five recent examples, among an increasing number:

- a report on Sudan by the United Nations Environment Programme (UNEP) which concludes that the 'impacts [of climate change] are closely linked to conflict in [Northern Darfur]' (see UNEP, 2007);
- an article by David Zhang and colleagues at the University of Hong Kong which argues that 'it was the oscillations of agricultural production brought about by long-term climate change that drove China's historical war-peace cycles' (see Zhang *et al.*, 2007);
- a seminar held at the Royal United Services Institute in London on 12 December 2007 which was entitled 'Weather of mass destruction: climate change as the "new" security problem'; the presentation (by Oli Brown of the International Institute for Sustainable Development) argued that 'the way we think about climate change may have to grow from a concern about environmental and economic damages to a recognition of the need for secure political systems that can weather the upcoming storm of adaptation to climate change';
- a number of speakers at the Bali climate-change conference on 3–14 December 2007 who emphasised the security implications of climate change, among them UNEP's executive director, Achim Steiner (see Howden 2007);
- a report by the NGO International Alert, written by Dan Smith and Janani Vivekananda, which claims to identify 'forty-six countries at risk of violent conflict and a further fifty-six facing a high risk of instability as a result of climate change' (see Smith and Vivekananda, 2007).

65 Hulme, M. (2007) Climate security: the new determinism *openDemocracy* 20 December www.opendemocracy.net/article/climate_change/the_new_determinism.

Old idea, new idiom

We now have climate as conflict, and weather as weapons. But such climatic determinism has had a long, and often discreditable, history. Climate has been viewed as the determinant of racial character, of intellectual vigour, of moral virtue and of civilisational superiority. The allure of a naive climatic determinism has seduced the Greeks (Herodotus), thinkers of the European Enlightenment (Montesquieu, David Hume, Buffon) and American geographers of the early twentieth century (Ellsworth Huntington, Ellen Churchill Semple). And it is now seducing those hard-nosed and most unsentimental of people … the military and their advisors.

The seduction has been underway for several years. In 2003, the United States defence department commissioned a study – *An Abrupt Climate Change Scenario and its Implications for US National Security*, written by Peter Schwartz and Doug Randall, and published in October that year – which presented a grim future of warring states and massive social disturbance as a result of climate change. In the era of 'war on terror', a new linguistic repertoire has been mobilised with which to describe climate change – as 'more serious even that the threat of terrorism' (Sir David King), as 'a weapon of mass destruction' (Sir John Houghton), as demanding a 'war on global warming to replace that on terror' (Stephen Hawking), and as 'the ticking clock' to replace the spectre of nuclear holocaust (John Ashton).

The British government, having clumsily framed Saddam Hussein's illusory weapons of mass destruction as a threat to global security, has now opened up a new security front in the United Nations. On 17 April 2007, the UN Security Council held an open debate at ministerial level on the relationship between climate change and international security. Their argument echoed the old deterministic one: climate change was a driver for conflict in the way that it exacerbated border disputes, encouraged mass migration, increased energy and resource shortages, and intensified social stresses and humanitarian crises. The aforementioned conflict in Darfur, western Sudan, became the exemplar for this line of reasoning: it is argued that drought, caused by anthropogenic climate change, lies behind the religious and ethnic confrontation which has seen more than 200,000 people killed since 2003. David Zhang's Chinese climate wars fit perfectly this new idiom.

The military turn

There are three reasons why the rise of this new climate-security discourse is worrying. First, the evidence for climate change triggering or worsening violent conflict is thin to vanishing. Such links are unsubstantiated by empirical evidence and serious academic study or, where cited, usually draw upon the second- or third-hand information and claims found in reports from think-tanks and advocacy agencies.

On the face of it, the empirical evidence is to the contrary. A study published in the 2007 issue of the *Journal of Peace Research* shows that since the end of the cold war the number of ongoing state-based armed conflicts has declined by a third. And, as summarised by Ragnhild Nordås from the Centre for the Study of Civil War in Oslo: 'While it is possible that climate change may lead to more conflict in the future, it has not so far caused a reversal of the current trend towards a more peaceful world.'

This finding builds on earlier researches, such as the first Human Security Report (HSR), part of a project based since May 2007 at the School for International Studies at Simon Fraser University. Published in October 2005, the report documents 'a dramatic, but largely unknown, decline in the number of wars, genocides and human rights abuse over the past decade'. The second HSR, published in December 2006, confirms this trend (see Paul Rogers: A world becoming more peaceful? 16 October 2005, *openDemocracy*).

The second worry about this line of reasoning is the intrinsic weakness of the old climatic determinist's position: it collapses the complexity of social, ethnic, political, economic, and cultural interactions into a one-dimensional narrative of cause and effect. Instead of Ellsworth Huntington's climatically-dictated hierarchy of superior races and civilisations – which just happened to favour the east-coast United States and England – we now have a convenient excuse for wars, violence, conflict and bigotry brought on by migration. ('It's the climate, stupid'). This is very reminiscent of the debate which went backward and forward for fifteen years in the 1970s and 1980s about climate change and desertification in the Sahel. Changes in climate were a very handy excuse for national governments in the Sahel, and for United Nations agencies such as the United Nations Development Programme (UNDP), wishing to deflect attention from poor governance, poor land-management regulations and unwise investments as the true reasons for deteriorating drylands.

The third reason for concern is the possibility of climate change being hijacked by the military agencies and interests of the world's more powerful nations. It becomes a power-grab by an instinctively hegemonic institution of society. The scientists, environmentalists, development experts and economists have all had their turn with climate change; now it is the military's moment. Even if the evidence for the destabilising role of climate in security terms was plentiful – and it is not, see above – it is doubtful that climate policies built on the premises of national and international military strategists will bring benefits to those most affected by climate change.

A reductive story

In view of these worries, the best we can say is the following. The constituents of both individual human security and collective national (and international) security are multifaceted. In the web of institutions, capital flows, relationships and narratives which affect such security it would be surprising if climate was entirely irrelevant. But we need more nuanced research and more complex sets of reasoning to put climate change into its proper place in the order of things.

What climate change means to us and means to the world is conditioned by what we do, by the way we govern, by the stories we tell. Presenting climate change as the ultimate security crisis is crudely deterministic, detached from the complexities of our world, and invites new and dangerous forms of military intervention. As we well know, military interventionists are not shy of using dubious claims in support of action.

The crude climatic determinism of Ellsworth Huntington was vanquished in the 20th century as the imperial ideologies of the British and their Anglophone cousins were exposed and thwarted. We must not allow a new form of determinism, expressed now in the quasi-militaristic language of security so beloved of the neo-conservatism of the post-cold-war era, to be its progeny.

24 A cultural history of climate

June 2010

A review of: 'A *Cultural History of Climate*' by Wolfgang Behringer for *Reviews in History*[66]

I received the invitation to review this book during the same week – 16–20 November 2009 – that over 1,000 emails to and from climate scientists in the Climatic Research Unit at my university found their way into the public domain. In the months since, climate science and climate scientists, and particularly these scientists who were concerned with reconstructing past climates, have been subject to phenomenal scrutiny. The so-called 'Climategate' has triggered a scientific controversy that will in due course play a central role in any cultural history of climate in the twenty-first century.

The author of A *Cultural History of Climate*, German historian Wolfgang Behringer, could hardly have foreseen this latest turn in the story of climate change. The book was originally published in German as *Kulturgeschichte des Klimas* in 2007, but has only recently been translated into English. Yet Behringer was prescient in using the so-called 'hockey-stick' graph – a graph showing estimated land temperature for the northern hemisphere over the last 1,000 years – as his opening framing device for the book. He observes that our efforts to reconstruct histories of physical climate can never be separated from the meanings that become attached to such reconstructions, and that these meanings emerge from particular political and cultural contexts. For Behringer, this key observation would seem to provide the rationale and motivation for his book. As he concludes on p.217: 'We cannot leave the "interpretation" of climate change to people ignorant of cultural history'.

In the 200 or so pages between these opening and closing remarks, Behringer offers an account of how changes in physical climates over 10,000 years have influenced human societies and how such changes have been understood by those societies. He is concerned to show not only the changeability of physical climate and the adaptiveness of societies to such change, but also how the ways people think about and make sense of climate and its variations – the 'behaviour' of climate we might say – are themselves mutable. Our present moment at the

---

66 Hulme, M. (2010) Book review of 'A Cultural History of Climate' by Wolfgang Behringer *Reviews in History* (review no. 925) http://www.history.ac.uk/reviews/review/925.

beginning of the twenty-first century offers a particularly powerful narrative about climate change, its causes and its consequences. Yet it is a narrative which, as well as being powerful, is sufficiently plastic to allow many different knowledge, policy and moral entrepreneurs to work with and exploit the idea of climate change in different ways. It is a plasticity that I explore in my own book *Why We Disagree About Climate Change* (Hulme, 2009).

Behringer wants to offer a deeper historical context to this interplay between the climatic and the cultural, between the physical fortunes of climate and the ways human societies think about what they observe happening around them. In so doing he follows in the footsteps of other climatic historians and historical geographers who have taken climate as their subject – historians such as Emmanuel Le Roy Ladurie, Hubert Lamb, Brian Fagan and Lucien Boia. The ground Behringer covers is therefore not unfamiliar. He starts with a broad outline of the world's evolving physical climate during the period of human evolution; he is interested in mid-Holocene climates and the emergence of human civilisations; he makes brief forays into Roman and medieval climates, including the obligatory stop in Greenland; and then he sojourns longer with the colder (European) climates of the early modern and Enlightenment centuries.

This is all very good, but we have heard a lot of it before. Where Behringer is perhaps at his most distinctive is in his narration of the cultural engagement of European societies with the cooling climate of the early-modern period, an era he has written about elsewhere especially on the subject of witchcraft and climate. And his account of the changing moral economy of European climate during the fifteenth to seventeenth centuries is pertinent to our current discourse about climate change and morality. For Behringer, the strong link between the 'little ice age' and witch persecutions came neither from the church nor from the state; it came 'from below' (p.132). It came from the populace through their search for accountability and meaning. What we are seeing at work today in our own society is a struggle between elitist and popular presentations of climate change and of its moral and political meanings.

Behringer's exploration of these issues in different historical contexts carries its own distinctive style which, one presumes, is well captured in this translation from the German. His is a less impressionistic study of the 'long Holocene summer' and the 'little ice age' than are Brian Fagan's parallel studies (Fagan, 2000; 2004) and is less analytic in approach than Jean Grove's exhaustive study of the 'little ice age' (Grove, 1988). It is less philosophical that Lucien Boia's historical account of climate and the human imagination (Boia, 2005). If it is less impressionistic, analytic and philosophical than these other studies, what is Behringer's characteristic style? It is perhaps most reminiscent of Hubert Lamb's classic book *Climate, History and the Modern World* (Lamb, 1982), which was published in 1982 and which left an indelible mark on me as a young geography student. Behringer offers less original scholarship than did Lamb (in his 1982 volume and in his other monumental studies), but with a rather straight-forward, unembellished style of writing Behringer brings the same sense of immediacy and authority as one gains when reading Lamb.

On the other hand because it adopts such a synoptic stance across such a broad reach of human history, his account suffers from being rather disjointed, lapsing at times into a string of historical climatic factoids packaged in their cultural correlates. What I therefore most missed in A *Cultural History of Climate* was a theoretical framework for thinking about the ways in which climates and cultures interact with and shape each other. It is a book lacking a big idea, or it is a book lacking at least an articulated and consistent perspective from which the reader can make sense of what is going on. Without such a framework in place, Behringer too frequently seems to be reading history through the lens of climate, a project which carries similar dangers as our contemporary obsession with trying to invent the future through the lens of climate (see below).

The author is too sophisticated a scholar to be lured too far down the line of climatic determinism, although at times he seems to get rather close to such reasoning. The Swiss, British and Scandinavian wars of the 1310s, for example, are implied by Behringer somehow (it is not clear how) to be linked to the cold winters and wet summers that have been suggested for that decade (pp.103–4). Elsewhere, however, he is willing to challenge the one-dimensional thinking of determinism, for example in refuting the supposed causal link between the 'worsening climate' of the early Middle Ages and the out-migration from the Roman Empire contributing to its decline (p.66). The difficulty here is one of scale and resolution. The more closely the history of an era, a period or a place is written, the more contingent and opaque becomes the role of any given weather phenomenon or climatic perturbation. The greater depth and resolution at which any historical event is examined, the more it emerges that human contingency dominates the direct physical effects of weather and climate. Only by staying at the macroscopic level of generalisation – at the synoptic scale – do cause-effect propositions about climate and society make any sense.

This methodological problem is linked to the difficulty of moving forensically between weather and climate. In most of these types of explorations of relationships between climate and society, documented weather events and extremes too easily become for authors symptomatic of 'sudden climate shifts' or an 'unpredictable change in climate'. This is a problem Behringer is alert to: 'For people living at the time, short-term changes [in climate] had greater importance than medium to long-term ones' (p.89) and yet most documented accounts of medieval and early modern Europe contain 'observations about the weather [rather] than about the climate' (p.89). Although at times acknowledged, I would like to see this difficulty examined more explicitly in a wider theoretical and methodological framework.

Otherwise, the danger by extension is what I have elsewhere called 'epistemological slippage' (Hulme, 2011). This is a danger we see in our present climate-obsessed imaginations. Because scientists make the claim that future climate can be predicted, the future starts to emerge in our imaginations as a future that is climate-shaped. The predictive claims of the Earth system scientists bring along in their wake a motley collection of environmental scientists, geographers, engineers and so on, who make their own 'predictions' of what the future human impacts of such putative climates will be – rather losing sight of the radical social, cultural, technological and political upheavals that await us over the next 50 to 100 years. Epistemological slippage occurs.

The same danger applies to our reading of the past. Because scientists make the claim that past physical climates can be reconstructed, the past begins to emerge in our imaginations as a past that is climate-shaped. It is harder to understand the prevailing social, cultural and political milieu of the sixteenth century than it is to see that rivers were flooded and glaciers enlarged. Epistemological slippage again can occur – as in Behringer's assertion that 'the fall of Mayan civilisation has been re-enacted in climate models' (p.71). I really would like to see how this has been done! The author therefore navigates between the twin dangers of determinism and epistemological slippage and without articulating a theoretical framework to keep all the pieces in place, there is a constant danger of allowing climate to intrude too far into his story.

Another methodological question emerges when one reflects on why these types of historical enquiries are undertaken. Are we interested in such accounts of changing climates and cultures because we are seeking to reveal through cultural artefacts and practices the physical properties of past climate – to contribute to a project of reconstructing climate 'as it really was'? Or are we intrigued because we are interested how the idea of climate – and hence too its perceived physical behaviour – is fashioned by culture? It is not always easy to know where Behringer stands on this.

The directionality of causation implied is very different. For example, Behringer argues for a 'strong link between the Little Ice Age and witch persecutions' in Europe (p.132). But did the physical climates of the seventeenth century make possible (explain?) the persecutions of witches or do the persecutions of witches in the seventeenth century tell us something about the prevailing physical climate? It is not clear which way Behringer is arguing. And this methodological difficulty can end up leading to circular reasoning. In his discussion of the decline of the Roman Empire, Behringer points to the abandonment of many settlements north of the Alps in the early Middle Ages. This, he argues, points to 'changed climatic conditions that made the cultural hiatus necessary' (p.67). What evidence is being used here to reveal changes in physical climates, as opposed to how changes in climatic conditions are being used to explain cultural change? These difficulties are not unique to *A Cultural History of Climate*; for example one can also find similar ambiguities and circular reasoning in some of Brian Fagan's work. But they do point to the need to develop a more sophisticated theoretical and interpretative framework for talking about climate and culture.

My reading of *A Cultural History of Climate* leaves me with one final question that needs further scrutiny. It is a question equally relevant to our creation of future climates as to our reading of past climates. How do different climatic indices gain their moral polarity? Throughout Behringer's book – and indeed more broadly in most historical accounts of the interactions between climates and societies – the notation 'warm + wet = optimum' and 'cold + dry = pessimum' applies. Thus paleoclimatologists talk of the optimum Neolithic climates of 6 to 8 kyr BP, while Behringer labels his section on Roman climates as the 'Roman

Optimum': 'It should be noted that climate historians consider [climatic] conditions to have been favourable at this time' (p.64). Favourable for what and for whom – for the Roman elites or for the Romans' slaves? And did the Romans themselves believe their climate was worsening?

In climate and history literature we therefore frequently have climates 'worsening' and 'improving' and climates becoming more or less conducive for human development. Is a cold climate a 'bad' climate? Is a warm climate a 'good' climate? Or is the only 'good' climate a stable and predictable one? And in other contexts we use the language of 'normal' and 'abnormal' climates (Hulme *et al.*, 2009), a phraseology which again reveals our normative judgements.

Such unreflexive and universalising tendencies to moralise climate have allowed our contemporary discourse about climate change to make the ultimate reductionist step: the signature of global climate is reduced to a single index of 'global temperature' and global warming becomes constructed as undesirable, if not dangerous. Yet this reverses the polarity of other discourses of climate change. European medieval warmth is usually presumed 'good' and later coolness is presumed 'bad' – this seems to be Behringer's position – while the pioneering scientists of the later nineteenth and early twentieth centuries theorising about carbon dioxide increases and climate change, generally regarded such human-induced global warming as a 'good' thing.

The relationship between climate and society is much more subtle and particular than is suggested by such crude moral indexing. As Behringer himself observes 'nature is not a moral system' (p.212) and how climate gains its retrospective (and prospective) moral polarity is always an exercise in power. This is nicely illustrated in the case of the Ming dynasty in China. Behringer suggests that the peasant insurrection of 1643 which ended the dynasty was catalysed by extreme weather and yet three centuries earlier a 'similar climate-induced crisis' (p.114) had brought the Ming dynasty to power. Whether one describes either of these climatic perturbations as 'good' or 'bad' depends on one's political allegiances.

There remains much work still to be done in gaining richer understandings of how the changing contours of climate – both changes in physical climate and changes in our imaginative ideas of climate – interact with cultural life around the world. We have far from exhausted investigations into how such ideas from different historical, geographical and contemporary cultures work with and against each other. *A Cultural History of Climate* is largely a cultural history of European climate, although Behringer occasionally visits non-European cultures from time-to-time. It would be good to see companion studies from outside the boundaries of Europe. Tim Sherratt and colleagues have attempted one such effort for Australia (Sherratt *et al.*, 2005) and William Meyer similarly for North America (Meyer, 2000), but neither of these extend further back than the early nineteenth century. But if our ideas of climate and climate change are indeed culturally inflected, then we need accounts that emerge from Brazil, China, India and Kenya before we can claim to have a world history of climate and culture. For example, I would like to know how the new Moghul rulers of India in the early sixteenth century understood and managed the variability of the Indian monsoon and how, as the Spanish set about establishing their New World empire at a similar time, the weather of central America was talked about.

We know that the weather and, by extension, our climate are important to us. And we know that this importance changes, just as we change. In reflecting on the place that climate has in our interior and exterior worlds we are too easily tempted to reduce climate to simple physical descriptive indices and/or to reduce the importance of climate to a simple determining role. Behringer's *A Cultural History of Climate* falls tantalisingly short of giving us the conceptual and analytical tools we need to resist these temptations, although he shows us why it matters that we do.

25 Learning to live with re-created climates

July 2010

An article published in
Nature and Culture[67]

> By seeing climate either as something to be idealised or as something to master, we fail to see what is happening to the world's climate. It is being reinvented as a novel entity, now co-produced between human and nonhuman actors. Rather than resist and lament the results of this new creative force, we must learn to live with them.

What we want from our climate

Why are we so frightened of a change in the climate? Why do we want our climates to stay the same, to offer us the same weather with which we have grown up? We welcome change and novelty in many other areas of our human experience: new technologies to captivate us, new places to visit, new people to love. Indeed, without experiencing some such change we feel stagnant, closed off, introverted. But we have decided – or been persuaded – that a change in the climate is dangerous. Dangerous for us, maybe. Dangerous for unknown and unborn others, definitely. We ask our politicians to act, we march on the streets in protest, we struggle to reduce our carbon footprints. We fear the clock is ticking to 30 November 2016 when the 100 months remaining before passing the tipping point expires.[68] We echo the call to 'stop climate chaos.' We demand the security and

---

67 Hulme, M. (2010) Learning to live with re-created climates *Nature and Culture* 5(2), 117–122. An early version of this essay appeared on the Copenhagen exhibition website *RETHINK: contemporary art and climate change* www.rethinkclimate.org/debat/rethink-nature/?show=bus in October, 2009.

68 The new economics foundation in the UK established the ticking clock to dramatise the notion of a tipping point in the Earth system which, they claimed, could be crossed one hundred months from August 2008 if greenhouse gas emissions continue unabated: (www.onehundredmonths.org). The tck tck tck campaign seeks to mobilise civil society and galvanise public opinion in support of transformational change and rapid action to save the planet from dangerous levels of climate change (http://tcktcktck.org).

the comfort of a stable climate. Some have written of a stable climate as a public good; as with trains that run on time and a stable currency. We want the seasons to return to what they were. We want the ice caps to grow back. We want to be free from the fear of flooding.

The myth of the lost Eden

But why do we want all this? Why was the 2009 Copenhagen climate conference proclaimed by some to be the last chance to save civilisation? It is because of the myth of Eden, a truth-laden picture-story, which embodies the idea of loss, lament, and a yearning for restoration. In this lament, climate becomes a symbol of the natural, of the wild, a manifestation of Nature that is (or should be) pure and pristine and is (or should be) beyond the reach of humans. Climate has become for us something that is fragile and needs protecting or saving, familiar goals that have fueled the Romantic, wilderness, and environmental movements of the West over two centuries. Changing global climate (inadvertently) through human actions becomes a threat to this last remaining remnant of the wild. British sociologist Steve Yearley (2006) has suggested that we are concerned about climate change not so much because of any substantive diminution of human or nonhuman welfare that might ensue, but because of the strong element of symbolism involved. We are concerned about anthropogenic climate change because our climate has come to symbolise the last stronghold of Nature, the final frontier resisting our encroachment. If we humans can alter the workings of global climate, there is nothing left to resist us.

The problem with rational climates

Anthropologists and environmental historians suggest that the idea of Nature as a separate category – something distinct from our interior worlds and therefore something that can be objectively studied by humans and hence physically damaged – emerged powerfully during the Western Enlightenment (e.g., Cruikshank, 2005). Those of us in these rationalistic societies have distanced ourselves from Nature. Yet it is a position that simultaneously, and paradoxically, opens up both destructive and romantic tendencies. By 'purifying' climate (Latour, 1993), by capturing and describing climate's essential character through numbers rather than through the imagination, we have allowed for projects both of domination and of re-enchantment. This is why we are so confused about climate change today (Hulme, 2009).

On the one hand, new techno-projects of climatic mastery are being proposed in which aerosols are pumped into the stratosphere and tiny mirrors injected into space. We want to shield ourselves from the sun. As with a thermostat in a room, we are inventing a thermostat for the planet. By approaching climate simply as a numerical index of global temperature to be regulated we allow ourselves to dream of such techno-solutions to our climate fears. As the mathematician von Neumann prophesied more than half a century ago: 'Intervention in atmospheric and climatic matters will come in a few decades and will unfold on a scale difficult to imagine at present ... what power over our environment, over all nature, is implied!' (von Neumann, 1955: 43). It is hard to imagine a project of greater human control and mastery than one that seeks to tame the sun and subjugate global climate to our fantasies and utopias (Fleming, 2010).

On the other hand, some of us react differently. Realising that our purifying tendencies have opened the way for both inadvertent and deliberate manipulation of climate, we seek forms of re-enchantment. This is the Edenic myth, our lament that by changing the climate, by losing wildness in one of the last untouched places, we are diminishing not just ourselves, but also something beyond ourselves. We are the poorer for it, as too maybe are the gods (Donner, 2007). This mythic position emphasizes the symbolic over the substantive.

It is a lament that underpins the deep ecology movement, which surfaces in some forms of eco-theology and which permeates, subliminally, contemporary discourse about climate change. The polar bear – that hackneyed icon of climate change – ends up not just worrying about its own survival, but also has to carry a huge additional weight of human nostalgia on its shoulders. American photographer Camille Seaman's haunting 2008 exhibition played to this lament:

> The Last Iceberg chronicles just a handful of the many thousands of icebergs that are currently headed to their end. I approach the images of icebergs as portraits of individuals, much like family photos of my ancestors. I seek a moment in their life in which they convey their unique personality, some connection to our own experience and a glimpse of their soul which endures.[69]

New hybrid climates

We cannot return to this mythical pristine climate of the past, neither by removing greenhouse gases from the atmosphere through collective policy action, nor by rebalancing the Earth's heat budget through some techno-fix. Too many other things are changing in our world for this to happen – and not just the sevenfold increase since the nineteenth century in the number of people living with us. Just as we cannot go back in time, neither can the Earth's climate system. There is no reverse gear.

Our climate then is no longer natural, nor can it be made so. There is no such thing as purely natural weather. The colour of the sky, the moistness of the air, the strength of the wind, even the warmth of the sun are all manipulated. Not artificial, but manipulated. By changing so substantially the composition of the world's atmosphere, we humans have not simply brought a new category of weather into being. The entire planetary system, which yields distinct weather at distinct times in distinct places, is itself now different. It is a hybrid system yielding hybrid weather. Whatever the weather outside your window today – calm or storm, warm

69 http://www.camilleseaman.com/Artist.asp?ArtistID=3258&Akey=WX679BJN.

or cold – is the result of this new coproduction between ourselves and the forces of Nature.

We may call this new construction a post-natural climate or a 'different paradigm of naturalness' (Yearley, 2006). But whatever we call it we cannot return to living with a simply natural climate. Climate is subject to the 'hot breath of humanity' – to use novelist Ian McEwan's memorable phrase – just as are forests, rivers and seas, and just as creatures, microbes, and nanoparticles. Climate and these other companions are all ours, both in creation and in relationship. We humans are now an actor in the unfolding story of climate's evolution, alongside the personal gods of the heavens and the impersonal dynamics of the oceans. We are now perhaps even the most important actor. The Enlightenment dream of separation, Latour's purification, is over. It is as irrelevant as it is impossible to find the invisible fault line between natural and artificial climate, between natural and artificial weather. We are now and forever the co-creators of our new hybrid climates.

Learning to rethink climate change

Should we fear our new role? Should we indeed be frightened of a change in climate? I think not. At least we need to learn to think not. We need to rethink climate change.

First, we must embrace the idea of novelty in our climates as much as in other forms of cultural innovation and in other areas of material co-production. Novel climates are neither good nor bad. They are novel, and we must find ways of imbuing them with meaning, value, and utility. Climate exists in the imagination as much as it exists 'out there' (Boia, 2005), and human societies have found many ingenious ways of living with the weather (Horn, 2007).

Second, we will not lose or deplete weather in the way that we may lose fish or deplete uranium. As climates change, the variety or volume of weather we experience will not be diminished, but the sequences and patterns of weather and the distribution of places where different weather occurs will alter. Rather than impoverishing our experience of weather in some way, the human generation presently alive has in fact experienced more weather than did any of our ancestors. We have become more mobile and our sensory encounters with climate are therefore more cosmopolitan (Hulme, 2010). Through new communication and digital media we encounter exotic climates vicariously in ways never before imagined. We are in fact the weather-rich generation.

Third, we must eradicate the notion of ever restabilising climate. Whether it is the imagined benign climates of Eden's past or the stabilised safe climates of the future controlled by a planetary thermostat, we delude ourselves to think that stability is the norm, or the achievable. In the dance between Nature's chaos and human intent it is change and variability that always prevails. Climate stability is a chimera and is best lost from our imaginations, and from our political and scientific vocabularies.

Neither lords nor serfs

We are not the lords of climate, exercising our vanities of mastery over Nature. We must not be lulled by the Earth system scientists, and their virtual climate miniatures, into thinking that we can rule the climate. Neither are we passive serfs, longing for a return to an unchanging golden climate of some imagined lost era. We must not allow our imaginations to re-enchant the climate with gods from the past. Rather we are intimate co-workers with Nature, jointly shaping our present and future climates and living somewhere in the creative spaces among imagination, knowledge, and ignorance.

26 A town called Bygdaby

10 April 2011

A review of: *'Living in Denial'* by Kari Marie Norgaard for *Nature Climate Change*[70]

The charge of 'denialist' has the potential to raise the temperature of any discussion of climate change by a few degrees. It is usually invoked by those who are frustrated either with criticisms of the trustworthiness of climate science or else with obfuscation about the desirability of taking action on climate change. It is also a claim that often triggers equally vehement claims of climate change 'alarmism', the result being a collapse of discussion into the simplistic binary trope of good versus evil.

It is therefore refreshing to read an account that treats climate change denial as an object of serious study. In *Living in Denial*, American academic Kari Marie Norgaard explores the sociological dimensions of denialism. She does so by moving the spotlight away from the overheated polemics of American or European media discourse, and instead turns it on a small rural Norwegian town that goes by the pseudonym of Bygdaby. This backwater community of 10,000 Norwegians becomes Norgaard's laboratory in which she explores the ways scientific evidence, personal experience, collective belief and cultural practice interact to lead to what she calls the social organisation of climate change denial.

Norgaard's approach is radically different to the trite moralising that characterises many of the exchanges that commonly take place on blogs about denialism and alarmism. As with all good systematic enquiries, she engages both with theory (in this case sociological and psychological) and with empirical evidence, allowing theory to shape evidence and evidence to re-shape theory. Her ethnographic evidence is gathered during a year – one that includes the mild and snow-poor winter of 2000 to 2001 – in which she lives as a member of Bygdaby town. She observes and participates in cultural activities such as sheep slaughtering and collective story-telling, and listens to the hopes and fears expressed in this unassuming community. She paints a picture of how a modest rural Norwegian society engages with the idea of climate change and how its people interpret it through their individual and collective world-views.

70 Hulme, M. (2011) A town called Bygdaby Book review of 'Living in Denial' *Nature Climate Change* 1(2), 83.

Through her direct observations, Norgaard helps us to better understand the cultural constraints that lead to quietism concerning climate change – the absence of social activism and public action. Norgaard attributes this lack of response to the phenomenon of socially organised denial, in other words the fact that information about climate science is known in the abstract, but is disconnected from political, social and private life.

Living in Denial adds to the small but rapidly growing body of anthropological and sociological work on human-induced climate change. Collectively, this work is starting to reveal how citizens in diverse cultures make sense of climate change for themselves, rather than simply imbibe what scientists say climate change is and means. Norgaard's study adds to this literature a rich and textured illustration of two important truths about how the idea of anthropogenic climate change works in the human world.

The first is that science alone cannot impose meaning on any physical phenomenon. Scientific evidence – whether about climate change or about the human genome – is always contextualised and interpreted through cultural filters. The meaning of a scientific fact is not for science to define. The second truth is that with our psychological and cultural heritage we find it very hard to engage imaginatively and emotionally with largely invisible and globally mediated risks such as anthropogenic climate change. In this respect, Norgaard's study is valuable for her deep emphasis on 'the feelings that people have about climate change and the ways in which these feelings shape social outcomes'.

Living in Denial is not for those who are looking for some secret key to unlock social action on climate change in the industrialised world. Norgaard has no time for the deficit model of communication in which people are bullied into action by sheer weight of information. Instead she offers an almost compassionate view of denialism as emerging from what Yale law professor Dan Kahan, and before him anthropologist Mary Douglas, has called the cultural cognition of risk. Norgaard moves the analysis of denialism to another level. The problem of climate change is not really about climate change at all; rather '[climate change] provides a window into a wholly new and profound aspect of the experience of modern life'. When engaging with the idea of anthropogenic climate change, people find new contradictions emerging between knowledge, values and actions – and they also find that there are no easy ways of resolving them.

Yet from this vantage point of understanding, Norgaard's own prognosis for climate change seems surprisingly parochial. Her call for a 'fierce return to the local' and for bottom-up community mobilisation seems inadequate for the task in hand. Although such responses may account for the community sensibilities and individual emotions Norgaard has astutely observed in Bygdaby, they leave untouched the much larger political and macroeconomic structures by which the lives of twenty-first-century humans are constrained.

One paradox of *Living in Denial* is that it reveals a distinctive local culture that seems resilient to the narrated threat of climate change. Cultural practices and collective beliefs in Bygdaby stabilise community life rather than unsettle it. They allow the social organisation of denial to emerge as a form of resistance to external global-scale challenges. This perspective challenges the positive valency that has recently been attached to the idea of resilience. Rather than being a desirable property of communities, cultural resilience may in fact become subversive by disabling radical forms of social and political change.

Here is where the real challenge of climate change rests, for denialists and activists alike: deciding who is culturally authorised to lead the charge for re-thinking and re-inventing social life in what is now inescapably a globalised and deeply interconnected world. It used to be kings and priests. Modernity then tried politicians and scientists. We now seem to be trying celebrities and bloggers. But who would the citizens of Bygdaby trust to lead them out of the land of slavery and denial?

Section five

Policy

Introduction

The articles in this section span more than 15 years of climate policy development, from the negotiations in 1997 leading to the Kyoto Protocol to the more recent adoption of '2 degrees' as the ostensible target of global climate policy. These items are selected to reveal my changing position in relation to international climate policy frameworks, a change which reflects the shifting geopolitical landscape over this period and my own views of what is politically achievable and desirable.

The first article (Chapter 27) was co-authored with Martin Parry and was published in *New Scientist* magazine during the first week of the Third Conference of the Parties (COP3) to the UN Framework Convention on Climate Change which took place in Kyoto in December 1997. Given the unlikelihood of tight emissions controls being agreed and implemented by politicians, and also given the deep inertia of the climate system, our aim was to draw attention to the importance of climate change adaptation. Although the Kyoto Protocol was negotiated during those 10 days in 1997, it was not ratified for a further eight years. Three years later in November 2000 – also for *New Scientist* – I wrote another essay in the lead-up to COP6 in The Hague. This article (Chapter 28) coincided with the formal opening of the Tyndall Centre for Climate Change Research (see also Chapter 15) and in it I framed three broad policy-relevant questions which required disciplinary research collaboration: What rates of climate change are dangerous? How can greenhouse gas emissions be reduced? And how can institutions adapt to changing climate risks? I foresaw the Tyndall Centre tackling all three.

When the Kyoto Protocol was eventually ratified in February 2005, attention was already focused on the G8 Summit to be hosted by Prime Minister Tony Blair at Gleneagles in July of that year. Climate change was one of the two major items on Blair's agenda, the other being global poverty especially in Africa. Chapter 29 is a review of Bjørn Lomborg's newly published book *Global Crises, Global Solutions*. Lomborg draws upon the results of an orchestrated cost-benefit analysis by some hand-picked leading economists to argue that attending directly to poverty reduction yielded a greater welfare benefit than investing equivalent amounts in climate mitigation. This was a controversial message in the months leading up to Gleneagles and in my review I point out some of the reasons why.

The value of Lomborg's analysis lay more in posing the right sort of question – 'what are our policy goals?' – rather than in offering a totally convincing answer to it.

The Gleneagles Summit, the entry into force of the Kyoto Protocol and the introduction of the EU emissions trading system, all in 2005, had failed to convince me that the policy framework for responding to climate change was appropriate. The next item (Chapter 30) is an unpublished speech which I delivered to a symposium organised in February 2007 by the Centre for Research in the Arts, Social Sciences and Humanities (CRASSH) at the University of Cambridge. I point out, somewhat rhetorically, the dismal achievements of climate policy to date. More importantly, however, I sketched out a form of thinking about climate change which I was later to develop in my book *Why We Disagree About Climate Change* and which was also reflected in The Hartwell Paper of 2010 (see Chapter 34). This drew upon the idea of 'clumsy solutions' and sought to reduce the complexity of the multilateral climate change policy framework into a series of more discrete, less ambitious, but more attainable policy goals.

Around this time, early in 2007, The Stern Review and the newly published Working Group I report of the IPCC's 4th Assessment Report were setting the agenda on public and policy debates about climate change. Chapter 31 is a short commentary on The Stern Review written for the *Bulletin of the British Ecological Society* (a publication which shows the wide interest generated by the Review). I pointed out the particular politics of the UK which had given birth and shape to The Stern Review and why this economic assessment would not be the last word on the subject.

A few months after my CRASSH speech, I picked up the theme of clumsy solutions again, this time in a key-note lecture presented to the Earth System Governance conference in May 2007 at the Free University of Amsterdam. This conference was convened by Professor Frank Biermann, who had been developing the new idea of Earth System governance. But as my unpublished talk shows (Chapter 32), I was as sceptical of this orchestrated form of planetary governance as I was of the targets-and-timetables approach of the Kyoto Protocol. Fragmented and localised policies were preferable to the hubris and hegemony of a single co-ordinated governance system. Frank Biermann also provides the link to the next essay included in this section (Chapter 33), an invited commentary on his proposal for a new UN Protocol for the Recognition, Protection and Resettlement of Climate Refugees. This was published in the US-magazine *Environment* shortly before the COP14 meeting in Poznan and in my commentary I suggested that such a Protocol dangerously over-simplified the relationship between climate, society and migration (see also Chapter 23).

The final denouement of the old multilateral targets-and-timetables policy regime took place in December 2009 during the COP15 meeting in Copenhagen. The weight of public and policy expectation had been extraordinary in the months leading up to COP15 and it was hardly a surprise that events took a very different turn to those hoped for by many. My reaction to COP15 is summarised in Chapter 34, an invited essay in a special issue of *Environment* magazine in May 2010 devoted to the aftermath of Copenhagen. This essay develops ideas previously offered in my CRASSH and Amsterdam speeches, but now given additional impetus by co-authorship of *The Hartwell Paper: a new direction for climate policy after the crash of 2009*, which appeared in the spring of 2010 (Prins et al., 2010). The emphasis on adaptation is consistent with my 1997 argument (see Chapter 27), even if the context in which I was now promoting it had radically changed.

The disillusionment amongst many climate change commentators and policy advocates which followed Copenhagen gave further impetus to the already looming idea of climate engineering: the idea of developing, testing and, if need be, deploying large-scale intervention technologies to manage the planet's heat budget. By the spring of 2010 there was a growing sense of the inevitable about geoengineering and this was marked by the Asilomar Conference on Climate Intervention Technologies which took place in California in March. My antagonism to this form of intervention, more specifically to solar radiation management, is expressed in Chapter 35, an essay for Yale University's environment blog, *Yale Environment 360*. In this essay I develop a scenario for the year 2028 through which I explore some of the many reasons why such interventions are likely to be much riskier than the greenhouse-gas induced climate change they are intended to neutralise.

The final item in this section (Chapter 36) is a critique of the '2 degrees' climate policy target which in recent years has become central in the international conversation about climate change. I wrote this short essay as a contribution to a book edited by Ottmar Edenhofer and colleagues in Germany, published in 2012 as *Climate Change, Justice and Sustainability* (Edenhofer et al., 2012). My contribution was intended as a provocation which was then evaluated by a series of commentators. Consistent with my earlier critique of the Kyoto Protocol and Earth System governance, I argue that the '2 degrees' target distracts from more important and more attainable policy goals.

27 Whistling in the dark

6 December 1997

Essay written with Martin Parry for
New Scientist magazine[71] and published
during the negotiation of the
Kyoto Protocol

Saving the world takes more than just one summit, say Mike
Hulme and Martin Parry

The conference of the UN Climate Convention now under way in Kyoto has
been portrayed as an 'all or nothing' opportunity to bring global climate under
more deliberate human control. Headlines such as 'The heat is on: Clinton's
policy dooms us all to a warmer world'[72] have thus appeared in recent weeks.

However, we must distinguish between three possible outcomes from Kyoto.
One would be an agreement that is so trivial, or even nonexistent, as to discredit
the Convention as an effective mechanism for bargaining around the global
climate table. Another would see the Convention survive essentially intact, with
agreements to some modest curbs of emissions in industrialised countries. But
such curbs will have little tangible benefit as far as averting climate change is
concerned. A third outcome would be one in which emissions controls were
agreed that translate into a significant slowing down of global warming – an
agreement which could be described as 'controlling the world's climate'.

Much of the discussion surrounding the conference seems to have assumed
that the third outcome is the ultimate objective of Kyoto. It is, however, an
idealist's dream, totally divorced from the realities of how the climate system
works. Global climate is not so easily controlled. The real debate after Kyoto will
not be whether the global climate is under control, but whether we still have a
credible Climate Convention.

71 Hulme, M. and Parry, M.L. (1997) Whistling in the dark *New Scientist* 6 December,
    p.51. A version of this essay also appeared a few days earlier, 2 December, in *The
    Guardian* newspaper.
72 *New Scientist* (1997), 1 November, p.5.

The difficulty of controlling global climate can be illustrated by taking several of the proposed emissions controls and running them through a simple climate model – one used by the Intergovernmental Panel on Climate Change (IPCC). Using a standard 'no-action' emissions scenario, the global warming by 2050 will be about 1.6°C over 1990 levels. For most of the credible targets so far proposed – the USA call for stabilisation by 2012, the call from small island states for a 20 per cent reduction in carbon dioxide by 2005, and the European Union's 15 per cent and Britain's 20 per cent reduction in greenhouse gases by 2010 – the rate of global warming falls to about 1.5°C by 2050 (to the nearest 0.1°C the three targets are indistinguishable). Only Brazil's proposal of a 30 per cent reduction in greenhouse gas emissions by 2020 brings global warming down by a further 0.1°C, to 1.4°C.

So, in hard global climate terms, Kyoto is about whether global climate warms by 1.6°C by 2050 or by 1.5°C. And we can take this exercise one step further by converting these calculations of future global warming into estimates of the number of people at risk from starvation. Using published crop and food trade models, we estimate that the difference between the 'no-action' case and, for example, the EU proposal, is 2 million fewer people at risk from starvation due to climate change (20 million will starve as a result of climate change rather than 22 million). This does not fit with the Kyoto rhetoric about 'saving the world'.

This perspective on Kyoto suggests a number of more realistic priorities. First, Kyoto must maintain the integrity of the Convention as a mechanism for global negotiations on climate control. The exact form of any agreement reached about the control of emissions is of secondary importance. Second, we need to realise that achieving the ultimate objective of the Convention – stabilising global climate at 'non-dangerous levels' – is a long haul in which the Rio Summit in 1992 was only the beginning and Kyoto is merely the next step.

Third, we need more thoroughgoing assessments of what constitutes dangerous climate change. Most of the Kyoto proposals are designed without reference either to the level of climate change to which we can adapt, or to the level of climate change that will cause significant damage. We are whistling in the dark. Despite the considerable achievements of the IPCC in its first two assessment reports, there is a need for a closer interaction between climate science and climate policy.

And finally, given the history of global greenhouse gas emissions and the inertia of the climate system, we are committed to a substantial amount of further global warming. This inertia in global climate demands creative, insightful and internationally just strategies to reduce our own and our children's vulnerability to climate change. Both mitigation and adaptation are appropriate responses to climate change, but adaptation may well prove the more necessary and achievable response in the near term.

28 Choice is all

4 November 2000

An essay written for *New Scientist* magazine,[73] coincidental with the official opening of the Tyndall Centre and just ahead of COP6 at The Hague

Spare a thought for climate scientists. On our backs rode the heroes of the 1992 Earth Summit and the 1997 Kyoto Protocol, which set international targets for cutting back greenhouse gas emissions. This month, governments will meet again, this time in The Hague, to finalise details of how the Kyoto Protocol will work. Many of the delegates gathering there will tell you that the climate scientists have completed their tasks. It's time, they'll say, for us to turn off our supercomputers, put away our temperature charts and go back to measuring the weather.

If only it was that simple.

Kyoto left climate scientists wiser to the ways of politics and international diplomacy. But the traffic all seems to have been one way. Politicians didn't come away with a reciprocal awareness of what climate science can still do for them as they turn their attention to setting policies to deal with global warming.

One of the issues high on the agenda at The Hague will be emissions trading. This is the idea that rich countries that emit more than their allowed quota of carbon dioxide or methane can buy the entitlement to do so from countries that emit less. But delegates will discuss this without actually knowing how much carbon or methane in the atmosphere poses real dangers.

Similarly, the US and some other countries would quite like to be allowed to plant forests in place of cutting greenhouse gas emissions. The idea here is that the trees will suck carbon from the atmosphere. But as any climate scientist will tell you, predicting trees' carbon uptake is a very imprecise science.

The Kyoto Protocol is built on the assumption that something, anything, must be done, and done quickly. It is an application of the precautionary principle: we may not have all the evidence to know precisely how climate change is harming the planet, but the risks associated with doing nothing outweigh the risks associated with taking action. But I think we can do better than this. In this important area of environmental policy, science really can deliver ... and here's how.

73 Hulme, M. (2000) Choice is all *New Scientist*, 4 November, 56–57.

There are three questions that lie at the heart of how we should respond to our changing climate. First: how can we tell what rate of climate change poses appreciable dangers to human health and to the environment? Second: can we control climate in a way that avoids such dangers? Third: if the answer to the second question is no, then can we shape our future world to accommodate the expected climate change? These are questions that scientists from a range of disciplines, from engineering to economics, can help to answer.

Many researchers believe that we need to limit global warming to no more than 2°C, and that it is dangerous for carbon dioxide concentrations to rise above 550 parts per million by volume, or twice the level before the Industrial Revolution. But not everyone agrees on this target, not least because it doesn't allow for the radically different effects that warming will have on different nations and regions of the world. A certain degree of warming could be catastrophic in some places while barely inconveniencing others.

So what exactly do we mean when we say a particular degree of warming is dangerous? Danger implies unacceptable risk. And to assess risk we need to know both the probabilities of particular outcomes and the consequences of those outcomes. The reality is that at the moment we don't know what these risks are. For example, is the collapse of the West Antarctic Ice Sheet (a low-probability, high-impact event) a greater risk to society than a 1°C warming (a high-probability, low-impact event)? Scientists can help to quantify these risks. Societies then need to arrive at a consensus on which risks they are willing to accept. Evaluating risk requires input from social scientists, as well as from climate scientists, physicists, and biologists engaging with society at large. Only then can we say with any confidence how big a temperature rise is too dangerous to tolerate.

That brings us to the second question: can we control climate sufficiently to prevent warming reaching this 'dangerous' level? It is the question that lies at the heart of the debate raging around the emissions targets set at Kyoto. These require the richer countries to reduce their greenhouse gas emissions by about 5 per cent before 2012.

But can this target, or any other, be reached? We don't know. We do know that ordinary people and businesses will have to change the way they live and make money. To find out whether they will do so needs research into the dynamics of technological change and research into the psychology of consumer behaviour. We need to answer questions such as why are Western societies so reluctant to consider nuclear power as an alternative to fossil fuels? And why are we so opposed to taxes on fuel, even when we know that some of the money will be spent on energy efficiency improvements? We need evidence-based policy.

More climate research is also critical for the third question: what happens if we cannot reduce the rate of climate change enough to avoid unacceptable risk? Can we adapt our institutions, our regulations, our behaviour to somehow accommodate more rapid climate change? Such questions have hardly been raised within the UN Climate Convention. Even talking about them is seen as defeatism, or at best as a diversion from the central issue of slowing down climate change.

In fact, identifying what 'insurance policies' we may need to manage climate change is an entirely rational response to the problem. Here again, a broad range of scientific expertise can help put the necessary debate on a sound footing.

To shape a global community that can evolve with a changing climate, we need to strengthen links between knowledge producers and the UN Climate Convention.

The real challenge for climate change science in the end is not to be able to predict future climate; rather, it is to give society the options to choose its own climate future.

29 Pie in the sky tops this G8's wish list

22 April 2005

A review of: 'Global crises, global solutions' edited by Bjørn Lomborg, written for *The Times Higher Education Supplement*[74]

> When eight economists considered how to spend $50bn, cost-benefit analysis was the name of the game. Mike Hulme laments a lack of sophistication

Have you ever won the jackpot? A £10 million Lotto rollover perhaps? Or have you at least dreamt of doing so? It is a game that most of us at some stage have played – what would you do if you won the Lotto jackpot? My own long-list of desirable investments would include a new sports pavilion for my local cricket club, improved cycle-track facilities for Norwich, a round-the-world tour with my family and a proportion for good causes. The way I would prioritise spending such a cash-limited windfall would include many factors – personal interest, ethical considerations, relationships and so on – by and large not easily reducible to economic cost-benefit analysis.

Bjørn Lomborg is also a dreamer. Yet, in addition, he is someone who relishes challenging orthodoxy, whether it be the political establishment in his campaigning Greenpeace days, the environmental movement in his 2001 book *The Sceptical Environmentalist*, and now all of us in his latest book, *Global Crises, Global Solutions*. This book is an account of the process established in 2003 by the Danish Environmental Assessment Institute – under Lomborg's direction – to determine how the world should spend an extra $50 billion, hypothetically made available by wealthy countries. This process resulted in the Copenhagen Consensus 2004, a statement from a panel of eight eminent economists that was announced last May. The panel

---

74 Hulme, M. (2005) *Pie in the sky tops this G8's wish list* Book review of 'Global crises, global solutions' (ed.) B. Lomborg in: *Times Higher Educational Supplement*, 22 April, pp.28–29.

concluded that the best thing to do with such a windfall would be to invest in new measures to prevent the spread of HIV/Aids – $27 billion would avert nearly 30 million new infections by 2010. Next, $12 billion should be spent on providing food supplements for the world's malnourished. In contrast, the panel claimed that investing in any of the three policy measures they considered that were aimed at tackling climate change would be a 'bad' investment, with costs likely to exceed benefits.

Like fantasising over lottery winnings, this is a great game to play and we can all join in. Lomborg, however, set up an elaborate and rigorous process with precise terms of reference, rules of engagement and eligibility criteria. Legitimate investments – those that might conceivably do the most 'good' outside wealthy nations – were long-listed by the institute committee; there were 32 challenges in all, reduced in the end to a short-list of ten challenges (or 'crises') for which promising 'solutions' were deemed to exist. To be eligible for playing the game you had to be a famous economist (ideally with a Nobel Prize to your name) with the time to spend five days in Copenhagen and to read more than 600 pages of briefing material beforehand. Out of the 6.5 billion people on our planet, only eight qualified.

Is this more than a game? Who should listen to the panel of eight elite economists? Who should read this book? The dilemma Lomborg has identified, and which the Copenhagen Consensus was established to address, is real – namely, on what basis are priorities set for tackling some of the world's besetting problems? As he correctly points out, nation-states have mechanisms for priority-setting in the investment of public funds and, in democratic nations, these priorities can be challenged every four or five years. And as the Lotto example mentioned earlier shows, each individual has a unique set of rules – idiosyncratic maybe, tacit certainly – for deciding where to spend scarce, or not so scarce, personal resources.

But what about the international system for investing public funds? What criteria are used to select between investing in what may be equally deserving causes? As Lomborg asks, is it simply those causes that are fortunate enough to have the most powerful or the most articulate or the most celebrated advocates that win? Do we simply have to accept pronouncements such as that offered by Sir David King, the Government's Chief Scientific Adviser, when he stated last year: 'In my view, climate change is the most severe problem we are facing today – more serious even than the threat of terrorism'.[75] I doubt his opinion is based on economic cost-benefit analysis.

This dilemma raises important and deeply political questions; indeed such questions are the very knotty problems that international politics, heads of state and foreign secretaries have to deal with week in, week out. My problem therefore with the Copenhagen Consensus – and hence with this book – is not that

75 King, D.A. (2004) Climate change science: adapt, mitigate or ignore? *Science* 303, 176–177.

Lomborg is too ambitious or bold in his project; my problem is that he is not ambitious or bold enough.

There are three main reasons for this. First, the focus of the consensus is the rather academic one of allocating a hypothetical resource. It ignores the more pressing issues of securing the resource in the first place and of implementing policy measures effectively to ensure its desired investment. Contrary to his claims about the consensus realists, in this regard it is Lomborg who is the dreamer. For example, finding ways of pushing developed nations that currently allocate 0.25 per cent of gross domestic product (about $69 billion a year) for official development assistance towards their target of 0.7 per cent ($193 billion) would seem a more important priority for influential economists with time on their hands. With regard to implementation, Lomborg and the consensus are also silent. The messy world of *realpolitik* has already reached powerful consensuses about tackling some of the world's crises – the Kyoto Protocol and the Millennium Development Goals come to mind. Is not the priority now to implement, and to monitor effectively, such agreements rather than for eight economists round a Copenhagen table to arrive at a rather (too) neat and artificial formulaic consensus?

Second, the focus of the consensus is too short term and lacks political vision. In this respect, contrary to the above criticism, the consensus process and outcome is too realist. Lomborg defends the short-term perspective – 'the next five years' – on the grounds that the panel wants to offer solutions that are tangible or provable. This might be sound for a certain class of problem-solving challenges, but I am not convinced it is appropriate for challenges such as climate change, arms proliferation and free trade. Here, actions taken in one electoral or economic cycle do not yield material or political reward for decades, if not for generations to come. No one who understands the climate system and the Kyoto Protocol would argue that the protocol, *sensu stricto*, will make a material difference to future world climate – but it is the political pathway the protocol opens up that is its most crucial feature. The lack of any coherent long-term political vision for the world means that at the heart of the Copenhagen Consensus is a rather mechanical and dispiriting utilitarianism – 'this is what we should do because the balance sheet will benefit'. This might be the world of *The Economist* (one of the sponsors of the Copenhagen Consensus), but it should not be the driving force for international politics, nor the essence of world leadership.

Notwithstanding the above two problems, the third reason why this exercise fails is that the boundaries of the consensus process are drawn too tightly. Lomborg's introduction to the process talks of the need to 'think outside the box' with regard to global problem-solving and the role of international agencies. Too often such agencies end up competing with one another or repeating solutions that have failed elsewhere. This is fair criticism, yet the very design of Lomborg's consensus exercise demolishes one set of boundaries – institutional – only to replace it with another set – disciplinary.

Features that are not explored or justified in the book are the framing of the process around the concept of cost-benefit analysis and the criterion for panel membership, namely, that only world-leading economists have the skills necessary for prioritising. This limited framing and execution leads to a rather stunted and unconvincing outcome, as reported by the panel members.

Why should economists have the privileged voice here? Who will represent those other factors that we know instinctively need to go into prioritisation exercises – planetary stewardship, cultural values, national identity and principle? The irony is that Lomborg recognises this limitation while nevertheless claiming legitimacy for the consensus outcome. For example, in excluding the $1,000 billion military expenditure from the prioritisation exercise he concedes that the lack of consensus on cost-benefit analysis as applied to international relations disqualifies this dimension of global investment from his programme. I am therefore surprised that he thinks there is consensus about cost-benefit analysis as applied to some of those global challenges he did feel confident in tackling – migration, climate change and corruption.

In the end, *Global Crises, Global Solutions* reports a brave and well-organised venture aimed at challenging our thinking about what matters most in the world. The greatest value I suspect are the ten chapters – one for each challenge – that introduce the reasoning behind global-scale policy intervention, defend a subset of specific policy measures and allow informed and detailed debate through the contribution of two opponents for each respective challenge. The ranking exercise, on the other hand – and of course most of the subsequent publicity was concerned with the ranks – was largely an exercise in rhetoric and publicity.

Global Crises, Global Solutions is aimed at each one of us and, more significantly, at those who influence international investments. But it fails in this more ambitious goal of changing the world of international politics and development economics. The Copenhagen Consensus ends up being rather reminiscent of a (admittedly sophisticated) high-school exercise in citizenry, one in which all thinking sixth-formers ought to participate. When repeated in 2008, I hope it moves from high school to university and embraces a more holistic view of what matters to humans.

30 A non-skeptical heresy: taking the science out of climate change

16 February 2007

An unpublished speech delivered at the seminar 'Debating climate change' organised by the Centre for Research in the Arts, Social Sciences and Humanities (CRASSH), University of Cambridge

For the fourth time in 17 years, the world's leading scientists who study the Earth's climate system have pronounced. In 1990, the First Assessment Report of the IPCC concluded that 'the unequivocal detection of the enhanced greenhouse effect from observations is not likely for a decade or more.' In 1996 the 'balance of evidence' suggested a discernible human influence on climate, while in 2001 most of the warming of the last 50 years was 'likely' due to greenhouse gas increases. And now, seventeen years after the first IPCC report, we now have the statement from the IPCC scientists that 'it is very likely' that most of the warming of the last 50 years has been caused by humans.

The Fourth Assessment Report of the IPCC is just the latest in a series of major reports, statements, campaigns and events of the last three years each of which have been heralded by many as marking *the* turning point in the fight against climate change. The siren voices thus brought forth have spoken from the worlds of science, economics, journalism and entertainment. They have been the prophets around which politicians, campaigners, celebrities, business leaders and scientists have congregated ... and on listening these apostles have amplified the message and transmitted it to the inner and outer reaches of our society. These megaphone voices have cajoled, scolded, bullied, preached, explained – pleaded even – with society to take heed, to repent, to change beliefs, behaviours and practises.

But the enlightenment has not arrived. Our hearts remain stubborn.

Some policy-tinkering has occurred here and there. Business has taken the plunge into carbon markets. Many millions of the masses have signed up for voluntary emissions reduction pledges. But emissions of greenhouse gases keep on rising and, of course, so does the world's temperature. Perhaps this particular way of framing climate change – as a Mega-problem awaiting a Mega-solution – has struggled down the wrong road.

Let me explain.

I will start with Hollywood. In May 2004, Roland Emmerich's film *The Day After Tomorrow* was released to the world offering the promise to some environmental campaigners of an induced sea-change in public opinion. Groups as diverse as Greenpeace and the Energy Savings Trust used Emmerich's apocalyptic vision of climate change to promote their cause.

Next it was the turn of science. In February 2005 the UK Government hosted the Exeter international science conference on 'Avoiding Dangerous Climate Change'. As a prelude to the July G8 Summit of that year, Tony Blair called on the scientists to identify what level of greenhouse gases in the atmosphere was self-evidently too much. For one week, the UK media struggled to keep pace with the stream of future climatic threats revealed by the (mostly English-speaking) scientists.

The journalists then had their turn. Following the Exeter Conference, *The Independent* newspaper started its climate change campaign which has been running ever since, joined briefly in September last year by a week of special climate change stories from … *The Sun* newspaper.

The climate system itself then stepped in with a demonstration of chaos and threw Hurricane Katrina at New Orleans in August 2005. And New Orleans was found wanting. For a while the public discourse in the USA was all about global warming, unfortunately deflated by the following year's hurricane season which was unduly pacific.

The BBC, not to be outdone by Hollywood, scientists or print journalists, commenced its own 'Climate chaos' season a year ago, inspired by the coalition of civic campaigning organisations convening under the slogan *Stop Climate Chaos*. David Attenborough lent his considerable presence to the season, with an encore just four weeks ago on BBC1.

And most recently it has been the turn of the economists led by Sir Nicholas Stern, under commission from the Prime Minister-in-waiting to soften up the international diplomatic community and to head off the new domestic political challenge on environmental policy being offered by David Cameron.

Many of the above reports and activities and campaigns have demonstrated professionalism of the very highest quality – film and TV producers, scientists, journalists and economists working at the tops of their professions and using all the tricks of their trades to entertain, explain, communicate and quantify. It has indeed been entertaining. And at times it has been dramatic.

And at one level it has been effective. We can now all talk about climate change. Taxi-drivers can be heard discussing what the carbon dioxide target should be in parts per million; the Confederation of British Industry have formed a climate change taskforce of business leaders; Leonardo di Caprio and Sienna Miller have endorsed the web-campaign *Global Cool*.

But where have these siren voices led us? What prospect of success do they offer? Emissions of carbon dioxide continue to rise – globally (by 16 per cent in the last decade), in the EU (since 1999, and now only 1 per cent below 1990 levels) and, perhaps most surprisingly, in the UK (after a fall in the 1990s, emissions are again on the rise). Few now believe the Kyoto targets will be met …

perhaps by Britain and Sweden if we are lucky, certainly not by the EU or the other signatories to Kyoto. And there are not many of us left who believe Kyoto Plus will look much like Kyoto. The explicit EU policy goal of limiting global warming to 2°C above pre-industrial (just 1.2°C above today's climate) – endorsed by the UK Government, by the Conservative Party and by most other environmental campaigners – has been fatally wounded. Initially, it was the compromise of an allowable 'overshoot' – temperatures could rise *above* 2°C as long as they eventually stabilised below this threshold. But this I suspect is only a first step towards an eventual relaxation towards a 3°C policy goal – as already hinted at by Sir David King, even if heavily criticised by those more tenacious for success.

The Emperor is exposed – he has no clothes.

Let me pause here in case I am misunderstood. The above account of the last three years is not written by a cynic. Nor by a sceptic. I have no doubt that climate is changing and that humans are now the dominant force in driving this change. We undoubtedly do have the unwanted, and rather unexpected, ability to shape the future of the world's climate. These changes warrant our closest attention and demand that we re-consider all our favourite development strategies in the light of this reality. I sincerely believe that the world would be a better and safer place with less carbon dependency in our energy system, that our atmosphere and oceans do not need the billions of tons of carbon dioxide we are pumping out year after year.

But we need to face up to reality. Nineteen years of IPCC assessments, fifteen years of the UNFCCC negotiations, ten years of activities inspired by the Kyoto Protocol (two of which have seen it in full force), successive rounds of G8 conferences in which climate has been agenda item number one – none of these sustained efforts has yielded the prize we sought. Earth system science has told us that the climate system is sluggish; we have now also discovered that the world's energy economy is sluggish and, no discovery this, that humans do not like being told what to do. As recognised recently by Sophie Hug – the *People and Planet* campaigner – 'we need to move from publicising the problem, which hasn't got us anywhere, to using the language of empowerment'.[76]

I conclude from all of the above that our current framing of climate change is wrong. We *have* been struggling down the wrong road.

Rather than seeing climate change as the Mega-problem to be 'solved', we need to identify and then separate out the different dimensions of the issue. At least two are easy to spot. The first is to develop a sustainable, secure and environmentally benign energy system. The second is to minimise the welfare costs that climate imposes on our societies (note, importantly, the cost that *climate* imposes, not climate change). These are two quite different problems operating on different time and space scales, with different communities of effective actors and with different available policy instruments. The politics required are also radically different.

---

76 Hug, S. (2005) What to do about climate change? *OpenDemocracy* 4 May 2005.

Sustainable energy and resilient societies in fact represent social and policy challenges that have been recognised since at least 1973 – the El Niño of that year brought home to us the huge impacts (especially in the Sahel) and costs of climate (again, note, climate not climate change), and the oil price shock of that same year demonstrated that oil was perhaps not the ideal fuel. This was well before these two almost unrelated concerns were unhelpfully stitched together in the official wording of the 1992 UN Framework Convention on Climate Change by well-meaning meteorologists and lawyers.

Since 1973 we seem to have made very limited progress on either of these challenges.

The delusion offered us by the current framing of climate change is that we can design, engineer and enforce a development pathway for world society for the next hundred years, if not beyond; the blind hope that with the aid of numerical models, quantitative economics (neo-classical or not) and global governance, we can achieve what has never before been achieved – designing, steering and then fine-tuning our global civilisation towards securing a normative long-term goal. Human beings are smart, but we are not this smart. And our faltering first steps towards such a hubristic goal should convince us of the fact.

Far better to focus on tractable, bounded problems where alignment of diverse political interests is achievable and where we can accept partial and clumsy solutions. If we are concerned about the impacts of climate change on future generations, let us first be concerned about the impacts of climate on the present generation. We *should* be concerned about the victims of hurricanes, floods and heatwaves, because we can do something to help them now. The science and technology to reduce many climate-related deaths and welfare losses already exists in our laboratories and in our best institutions of government. Yes, the deep structural crevasses of social inequality and injustice will be revealed by such actions and will need to be bridged, but this is exactly the essence of what we have to tackle in any case. We cannot be serious about climate change tomorrow if we are not serious about social injustice and chronic poverty today.

And then on energy. The real issue here is how to forge coalitions of the willing around mutually beneficial actions. The Chinese understand this. Energy to them is about efficiency, security, affordability, sustainability *and* environmental impact. It is not about climate change. We should not artificially force these various dimensions of energy into the straight-jacket of contentious emissions reduction targets. Nor should we burden such willing coalitions with having to find a way of connecting any agreed set of energy goals or actions with the different concerns of development, social justice and climate change adaptation. From a different starting point – from the position of revealing multiple, diverse yet synergistic goals – a stronger drive towards technological and behavioural transformation of our energy system is possible, even some may say inevitable.

But in the end society cannot be brow-beaten by science into such a transformation. Science – least of all Earth system science – cannot simply speak truth to power and all will be well. Science cannot tell us what is 'self-evident', to use Tony Blair's phrase. No number of IPCC reports, Exeter Conferences, World

Bank economists or Hollywood movies will do the job. Less likely still is that society will be re-fashioned through the use of terroristic language of a new environmental inquisition. As Ian McEwan has commented, 'It is tempting to embrace with enthusiasm the latest bleak scenario because it fits our mood. It would be self-defeating if the environmental movement degenerated into a religion of gloomy faith'.[77] We do not need to alarm ourselves with clocks of doom.

No. We need to let society take ownership of climate change away from the scientists and the UN bureaucrats. For too long we have been bewitched by the hubristic claims that the models of our scientists can define our future. Instead, we need a billion voices to speak. Instead of the prophets of the apocalypse, we need to use our knowledge of climate change to inspire the future, not be fearful of it. Climate change is an idea to be embraced, not a Mega-problem that can be solved.

The idea of climate change can help us solve the real problems that lie elsewhere in our world – poverty, social exclusion, corruption, infectious disease, affordable and secure energy for all.

77 McEwan, I. (2005) Let's talk about climate change *openDemocracy* 21 April.

31 The limits of the *Stern Review* for climate change policy-making

March 2007

A commentary on the *Stern Review* published in the *Bulletin of the British Ecological Society*[78]

The *Stern Review on the Economics of Climate Change* was published by the UK Treasury on 30 October 2006. It was announced to the world in a carefully orchestrated media campaign, spearheaded by the British Prime Minister and the Chancellor, in the week before five thousand climate change negotiators and actors congregated in Nairobi for the 12th Session of the Conference of the Parties to the UN Framework Convention on Climate Change.

The timing of the report and the authority the two most senior figures in British politics endowed it with was not accidental. Nor was the choice of the lead author, Sir Nicholas Stern, a former chief economist at the World Bank. The review therefore carried the imprimateur (even if by association) of that most establishment of financial institutions and, by being commissioned by the Treasury, the approval of another one. The *Stern Review* is a policy document, written by a team of civil servants, speaking to a very specific policy audience and it must be understood on those terms. It also draws upon a large mountain of submitted evidence from independent researchers across the UK and many countries further afield. The report therefore has cache in three of the circles of institutionalised power – politics, economics and science.

The report is undoubtedly impressive. Its 700 pages allow it to scan, sift and synthesise the scientific evidence for present and prospective climate change, assess what it is believed climate change might mean for society in the future, undertake a unique analytical exercise in attaching specific costs to these impacts, and survey and advocate a portfolio of policy measures that are worthy of consideration – some national, others with a more international flavour. The key message from the *Stern Review* is that the costs to society of doing nothing about climate change will eventually greatly exceed the shorter-term costs of re-directing the global economy onto a low carbon trajectory.

This message is spoken to two audiences. Nationally, climate change policy is now a crucial part of the actively contested territory between the three main

78 Hulme, M. (2007) The limits of the Stern Review for climate change policy-making *Bulletin of the British Ecological Society* March issue, pp.20–21.

political parties. The Liberal Democrats and the Conservatives are both seeking to out-flank Labour on this battleground. The *Stern Review* makes a strong statement legitimising new policy measures on climate change which the Government may want to introduce.

More importantly, the *Stern Review* also speaks to an international audience, as evidenced by the global tour of capital cities undertaken by Sir Nicholas Stern in the weeks and months since publication. The key message of Stern – action on climate change is needed to guarantee economic security – speaks into the world of international diplomacy which is struggling to come to terms with the challenges presented by climate change. The Kyoto Protocol expires in five years time and the nature of international co-operation on climate change beyond 2012 remains unclear.

The *Stern Review* is a megaphone voice from London speaking into the clouds of hesitation, deviation and prevarication which surround current international negotiations.

But it is a very British voice. By this I mean that no other nation on Earth could at the present time have delivered such a bold and pedagogic message about climate change. The *Stern Review* resonates well within the domestic political landscape in Britain, but perhaps nowhere else. The language of the *Stern Review* is increasingly the language of public discourse in Britain about climate change – where else would one find taxi drivers talking about whether 450 or 550 parts per million of carbon dioxide in the atmosphere was desirable? It draws heavily upon the UK science base, much more so than upon the scientific assessments of the IPCC. And finally the *Stern Review* fits perfectly with Britain's current diplomatic campaign of showing international leadership on climate change; it is an exercise in demonstrating British hegemony over climate change diplomacy. Put another way, a review of the economics of climate change commissioned by the Ministry of Finance in Beijing or Djkarta would look and feel very different to the *Stern Review*.

None of the above analysis is intended to argue that the *Stern Review* is an artificial or flimsy construction. It is just the opposite – it is a deliberate and substantial contribution to the debate. But how effective will it be and what difference will it make? In ten years time, will we be able to look back and analyse a pre-Stern and post-Stern discourse about climate change, or see 2006 as marking a break-point in climate policy?

I suspect not. To look for the reasons one need do no more than re-wind the clock to 1998 and the publication of the proceedings of the largest co-ordinated exercise yet undertaken by social scientists into examining the implications of climate change for human choice (Rayner and Malone, 1998). A self-proclaimed 'complement' to the United Nation's IPCC, this five-year assessment delivered ten suggestions for policymakers in regard to climate change. They deserve wider visibility and recognition. To understand the limits of the *Stern Review* let me mention just three of these ten suggestions, all of which emerged from an extended examination of knowledge emerging from the social sciences (and anthropogenic climate change after all has emerged from society, not from nature):

- 'Recognise that for climate policy-making institutional limits to global sustainability are at least as important as environmental limits'. The *Stern Review* has very little to say about new institutional arrangements commensurate with the nature of climate change decision-making. The barriers to effective action on climate change are not incomplete science or uncertain analysis, but the inertia of collective decision-making across unaligned or even orthogonal institutions.
- 'Employ the full range of analytic perspectives and decision aids from the natural and social sciences and humanities in climate change policymaking'. The *Stern Review* remains dominated by natural science and macro-economic perspectives on decision-making and although some concession to the role of values and ethics is made in the review, the values and ethical judgements made are pronounced rather than negotiated.
- 'Direct resources to identifying vulnerability and promoting resilience, especially where the impacts [of climate change] will be largest'. The *Stern Review* continues to place emphasis on linear goal-setting and implementation; a more strategic approach is to focus on measures that promote societal resilience and opportunities for strategic switching, informed by regional and local perspectives.

Climate change is not a problem awaiting a solution. It is categorically different to the depletion of the ozone layer over Antarctica where one could isolate the cause and effect chain and design a decisive intervention through a small number of actors to remedy the problem. Climate change is a phenomenon embedded in almost the entire diversity and geographical spread of human activity, enterprise and fulfilment, it emerges from the sense of identity and purpose we have (unfortunately) created for ourselves.

As the late [American political scientist] Stephen Meyer stated eloquently in his parting monograph *The End of the Wild* just days before he died, our efforts to protect our companion species on this Earth should, 'like the Ten Commandments, remind us who we could be' (Meyer, 2006: 88). As with biodiversity, so with climate. If we continue to measure 'progress' in the perverse way we do, if we continue to tolerate gross inequities in our contemporary social world, or if we continue to discount the value of the natural world, we will need to get adept at living in a world with a continually warming climate. No remedy will be in sight.

32 Setting goals for global climate governance: an inter-disciplinary perspective

24 May 2007

An unpublished speech presented at the conference on 'Earth System Governance: theories and strategies for sustainability' in the session 'Architectures of Earth System Governance – climate architectures after 2012', Free University of Amsterdam

Anthropogenic climate change is a pronouncement, although one we have realised only slowly, that the intellectual divide between Nature and Culture is redundant. These distinct categories have no bearing on reality. The era of the Anthropocene started before we had even invented it. By the time Paul Crutzen coined the term at the turn of the new millennium, we had already well passed the point when our collective footprint on the planet, and especially upon the atmosphere, was pressing so hard as to change the functioning of our Earth system.

As the series of IPCC assessments from 1990 onwards has made clear, not least the bold pronouncements about detection and attribution made by IPCC Working Group 1 in Paris last February, the defining physical transformation of the Anthropocene is now occurring above our heads. Our weather is semi-artificial, influenced increasingly by the changes we are imposing on the global atmosphere. It is becoming harder, not that it was ever easy, to separate Acts of God from Acts of Man. The insurance industry is having to re-think its language. Our climate is, to borrow an expression, a co-construction of Nature and Culture.

Since humans are now an active agent in shaping the climatic future we have inevitably had to face the question ... should our climate agency be inadvertent or advertent, purposeless or planned, a mere by-product of our pursuit of tangential individual and collective goals or a consequence of a deliberated and hard-won global strategy? Do we really want to share with Nature this burden of climate design?

Well, this decision was in fact taken by the world 15 years ago, in Rio in 1992. With the signing of the UNFCCC, 180 nations committed themselves to the task of climate control. We are therefore willing, keen even, to shoulder this task. But the extent to which we *can* consciously guide (dare we say manage?) global climate remains unknown.

The idea of controlling the weather has a long pedigree about which a fascinating story can be told, but one too long for this afternoon. Some of the grandest schemes for weather control have emerged from both wings of the ideological spectrum – the American frontiersman thinking of the nineteenth century and Soviet-inspired state control in the twentieth century. We also increasingly control the micro-climates in our cars, our homes and our buildings. Sophisticated engineering systems secure for their inhabitants a desired (and alterable) level of comfort. 'Climate-control' is now a standard selling point for all new cars. And where our habitats are open rather than closed – and where meso-scale climates cannot be engineered – we construct and reconstruct whole cities (like Phoenix or Las Vegas) or whole countries (like the Netherlands) to secure our survival in the face of uncontrollable aridity or threatening seas.

But to return to global climate control … we have taken on the task but where do we set the thermostat? What is the comfort zone for global climate – keeping it within the bounds of the last generation (±0.2°C), the last millennium (±0.5°C), or the Holocene (±1°C)? Are we brave enough, or foolish enough, to push the thermostat up by 2°C or more? What should our designer climate look like? What is the ultimate goal of our ambitious attempts at climate control?

A controlled system needs a controlling agent or agency … a master-mind perhaps, or a governor. So whose hand is on the control panel? Some may claim to see God's hand in control. Or maybe it is the 'invisible hand' of Adam Smith's market – even if it does have a 'green thumb' as Al Gore would like to think. Can we do better than the neo-liberal ideology of the market-place?

Many people think we can do better by creating a new global polity – the political hardware to enact global governance and policy. Hence the notion, I guess, of Frank Biermann's Earth System Governance and his four principles of: credibility, stability, adaptiveness and inclusiveness. This idea forms the very motive and rationale for this whole conference.

Now I don't want to focus on the future of Earth System Governance – the design and operation of such a system. Others are better qualified than I. But I do want to make some remarks on the ultimate goals of global climate control and what they signify to me. Of course the classic formulation of the goal remains Article 2 of the UNFCCC: 'The ultimate objective of this Convention is … to achieve stabilisation of greenhouse gas concentrations in the atmosphere at a level that would prevent dangerous interference with the climate system. Such a level should be achieved within a timeframe sufficient to allow ecosystems to adapt naturally to climate change, to ensure that food production is not threatened and to enable economic development to proceed in a sustainable manner.'

In the fifteen years that Article 2 has been on the negotiating table, the goals of climate policy have been framed in a number of different ways … as emissions, concentration or temperature targets; they have been proposed or implemented at a number of different scales … community, state, nation, region or world; and they have been mapped out at different distances into the future ... 2012, 2020, 2030, 2050. All of these framings have been loosely inspired by the 1992 notion of avoiding 'dangerous anthropogenic interference' with the climate system. Here are my seven different categories of goal-setting:

- technology goals – e.g. the Asia-Pacific Partnership: development, deployment, transfer of low carbon technology;
- carbon intensity/GDP – e.g. the USA to reduce intensity by 18 per cent by 2012;
- relative reduction in absolute carbon emissions – e.g. 60 per cent (or 90 per cent?) reduction by 2050 for UK; 20 percent (or 30 per cent?) by the EU; 25 per cent by 2020 by California;
- carbon or greenhouse gas concentration targets – e.g. 450–550 ppmv (Stern Review); or 380–450 ppm (Danny Harvey's recent suggestion[79] to avoid dangerous anthropogenic interference);
- global temperature – e.g. to ensure warming is no more than 2°C above the nineteenth-century level (the EU);
- biogeophysical – e.g. to avoid tipping points in the Earth System (melting of the Greenland Ice Sheet);
- social goals – e.g. the eight Millennium Development Goals: poverty, hunger, child mortality, infectious disease, gender equality, etc.

The sequencing of this list is important – it starts with society and ends with society.

And I want to suggest that these climate policy goals are always a proxy for other things – for maintaining ecosystem integrity, economic security and food production in the case of Article 2 – or for achieving or enhancing poverty reduction, trade liberalisation, national security, sustainable consumption, climate protection (by which I mean climate risk management), etc., as revealed by the various discourses that exist around global climate policy. In other words, these other things are subsumed within the head-lined goals of climate policy. And they are subsumed by construction. Climate can't be controlled without these other elements of our political, energy and welfare economies being reformed and re-structured, without these other goals either being met (or maybe missed).

These other underlying realities should therefore be exposed and accommodated in the processes of negotiation. How serious are we about the elimination of trade subsidies, about a commitment to open borders, about the eradication of hunger? Is it global or national security that we really care about? And it suggests to me that rather than trying to steer a complex climate system to a precisely (or even

---

79 Harvey, L.D.D. (2007) Allowable CO_2 concentrations under the United Nations Framework. Convention on Climate Change as a function of the climate sensitivity probability distribution function *Environmental Research Letters* 2 014001 doi:10.1088/1748–9326/2/1/014001.

loosely) defined outcome, these realities require a diversity of so-called 'climate goals' to co-exist in any governance regime. When dealing with such complexity we are better off focusing on constraining the inputs to the system rather than debating and predicting the chimerical outcomes.

Hence technology goals and social goals should be central to our thinking.

As the policy debate over fragmented versus universal governance regimes starts leaning in one direction – towards the fragmentation of actors, policies and institutions – so too should the debate over the goals of climate policy lean in this same fragmented direction. The paradox of our now co-produced and globalised economy and climate is that we can only hope to exercise any conscious control over such hybrid entities through fragmented and localised policies. Bottom-up interventions rather than top-down ones; pluralist clumsiness over hegemonic hierarchy.

I suggest that our fixation on global temperature, emissions reduction and concentration stabilisation targets, inspired by our lock-in to the Framework Convention, has placed a straightjacket on our thinking about the goals of Earth System Governance. It is hubris to think that we can macro-manage our way to a designer climate with an uncooperative Nature.

But we *can* hold out the possibility, at least the possibility, of micro-managing some of the constitutive and regional elements of our emergent global climate economy. But how Culture and Nature will then co-conspire to construct our future climate will probably surprise us.

33 Climate refugees: cause for a new agreement?

November 2008

A commentary on 'Protecting climate refugees: the case for a global protocol' by Frank Biermann and Ingrid Boas, published in *Environment* magazine[80]

With their proposed Protocol for the Recognition, Protection, and Resettlement of Climate Refugees, Frank Biermann and Ingrid Boas[81] make a bold and provocative suggestion, one of a number of political responses that are being floated as the world moves toward designing a new post-2012 architecture for managing climate change. Putatively a new protocol under the supervision of the UN Framework Convention on Climate Change, its goal would be to enable nation-states to manage proactively the resettlement of people who may face displacement due to climate change.

I see three significant flaws with the proposed protocol: the category of 'climate refugee' is essentially underdetermined; it adopts a rather static view of climate-society relationships; and it is open to charges of carrying a neo-colonial ideology, which guarantees it will meet political resistance.

For the protocol to be operational, it is necessary to clearly define who does and does not fall under the designation of 'climate refugee'. The term implies a mono-causality about the reasons for migration that just does not exist in reality. The decision to migrate is always a result of multiple interactions related to economic, political, environmental, and social factors.[82] Even in the case of Pacific Island states such as Tuvalu, sea-level rise is rarely the decisive factor behind observed population movements,[83] and Santa Clara University professors

80 Hulme, M. (2008) Climate refugees: cause for a new agreement? Commentary on: 'Climate refugees: protecting the future victims of global warming' by Biermann, F. and Boas, I. *Environment* 50(6), November/December, 50–54.
81 Biermann, F. and Boas, I. (2008) Protecting climate refugees: the case for a Global Protocol *Environment* 50(6), November/December, 8–16.
82 See for example Black, R. (2001) *Environmental refugees: myth or reality?* UN High Commissioner for Refugees Working Paper No. 34, University of Sussex, UK.
83 See Connell, J. (2003) Losing ground? Tuvalu, the greenhouse effect and the garbage can *Asia-Pacific View Point* 44, 89–107. Connell argues that economic aspirations and greater social mobility are inextricably bound up in the reasons for migration.

Michael Kevane and Leslie Gray have recently shown that the widely claimed climate-induced refugees in Darfur are nothing of the sort.[84] One is also reminded of Nobel Laureate Amartya Sen's claim that there has never been serious famine – nor associated migrations – in a country with a democratic government and a free press.[85]

Biermann and Boas side-step the problem of assigning the climate refugee category to individuals by proposing that entire communities or population groups are so designated by the protocol's executive committee. They suggest such designations should ideally be a pre-emptive move years or even decades before the prospective critical change in climate or sea level occurs. This is certainly one way of inflating the numbers of those to be considered climate refugees – numbers which have been critiqued by many.[86] This international committee thus could determine the fate of millions. Not only must these committee members discern, amid the enduringly fuzzy science, which habitats climate change will make unviable and by what approximate year this will occur, they also will have the even more problematic task of determining which areas 'are deemed as being too difficult to protect [through adaptation] in the long-term.' Adaptation is not a technical process to be determined or imposed by some distant UN committee; it is a social dynamic of change in which multiple values and power relations are at work.[87]

A second concern regards the relationship between climate and society implied by the proposed operation of the protocol. For example, Biermann and Boas explicitly state that once categorised as climate refugees, population groups must be treated as 'permanent immigrants to the regions or countries that accept them … [they] cannot return to their homes.' This implies a frozen view of reality – once an area becomes uninhabitable it always will remain uninhabitable. Yet we know from accounts of earlier migrations in climatically stressed regions such as the Sahel in the 1970s and 1980s that migration is often a temporary response to

84 Kevane, M. and Gray, L. (2008) Darfur: rainfall and conflict *Environmental Research Letters* 3: doi:10.1088/1748–9326/3/3/034006.
85 Sen, A. (1994) The political economy of hunger in: Serageldin, I. and Landell-Mills, P. (eds) *Overcoming Global Hunger, Environmentally Sustainable Development Proceedings Series No. 3* Washington, DC: World Bank, 88.
86 For example: 'One should be cautious when dealing with the estimations of numbers of "climate refugees" since there is not one common definition and the names and numbers are coloured by different discourse and agendas (such as the environmentalists, security, protection, etc).' Kolmannskog, V. O. (2008) *Future Floods of Refugees: A Comment on Climate Change, Conflict and Forced Migration* Oslo: Norwegian Refugee Council, p.10. The estimate of 150–200 million environmental refugees (from Myers, N. and Kent, J. (1995) *Environmental Exodus: An Emergent Crisis in the Global Arena* Washington, DC: Climate Institute, 1995, p.149) remains an untested and unverifiable number, as Biermann and Boas accept.
87 See discussion in M. Hulme *et al.* (2007) Limits and Barriers to Adaptation: Four Propositions, Tyndall Centre Briefing Note No. 20 Norwich, UK: Tyndall Centre, 2007.

environmental stresses.[88] We also know that habitability is deeply contingent – think cities such as Phoenix or Amsterdam.

The imposition of irreversibility and permanency as a condition of categorisation places too great a burden on being able to distinguish between human-related climate change (and sea-level rise), which is difficult to reverse, and natural, annual, or decadal climate variability, which by definition is reversible. This challenge to scientific knowledge is especially acute for all rainfall-related stresses since for most tropical regions, we continue to have little idea about the stability of local rainfall signals of anthropogenic warming.[89] Migrations linked to storms, drought, and famine are particularly subject to this ambiguity – an ambiguity far more intractable than Biermann and Boas allow for in their optimistic claim about the 'broad predictability of climate change impacts.'

My third concern runs deeper still and engages with the new geopolitics of climate change. Establishing a protocol that would be supervised by an international executive committee would open up a new front in the emerging debate about green neo-colonialism. New moves to establish international payments to tropical nations for preserving swathes of rainforest are also subject to this same critique of global environmental protection being used as an extension of the hegemony of international financial and political interests.[90] In whose interests therefore is the new refugee discourse (and protocol) being developed? A recent report from the Norwegian Refugee Council alerts us to the dangers: 'A fundamental critique is found in the context of North-South discourse where "environmental security" is seen as a colonisation of the environmental problems, suggesting that the underdeveloped South poses a physical threat to the prosperous North ... th[is] security discourse can serve to make new areas relevant for military considerations and promote repressive tendencies'.[91]

Furthermore, Biermann and Boas's protocol adopts a paternalistic and centralising approach to climate-related migration and resettlement. It is a long way removed from the participatory citizen-based dialogues between community, government, and stakeholders currently under way in countries such as the United Kingdom, where compensation for property and livelihood loss due to the

88 See, for example, M. Mortimore and W. M. Adams, *Working the Sahel: Environment and Society in Northern Nigeria* (London: Routledge, 1999), 226.

89 For example, see the table in J.H. Christensen and B. Hewtison (eds), 'Regional Climate Projections,' in Intergovernmental Panel on Climate Change (IPCC), *Climate Change 2007: The Physical Science Basis, Working Group I Contribution to the Fourth Assessment Report of the IPCC* (Cambridge, UK: Cambridge University Press, 2007), 847–940. This shows that for many tropical and subtropical regions, great ambiguity remains about what the long-term anthropogenic change in rainfall will be.

90 See the article, and subsequent responses, B. Jagdeo, 'Why the West Should Put Money in the Trees,' *BBC News Online*, 8 September 2008.

91 Kolmannskog, note 6, pp. 9–10. See also S. Dalby, 'Environmental Change and Human Security,' *Isuma – Canadian Journal of Policy Research* 3, no. 2 (2002): 71–79.

encroaching sea is a live issue.[92] A refugee regime as suggested here may only be viable in authoritative and centralised societies where the voices of citizens are rarely heard.

I remain unconvinced about the need or viability of such a protocol. Climate and development are embedded evolutionary processes, and both have dynamics that should not be artificially reduced to simple cause and effect, least of all if so doing opens the way for powerful vested interests to control personal and community development. The consequences of climate change and variability for human well-being, development, and migration are best handled within existing and evolving development and adaptation discourses and practices.

92 See, for example, J. Milligan and T. O'Riordan, 'Governance for Sustainable Coastal Futures,' *Coastal Management Journal* 35 (2007): 499–509.

34 Moving beyond climate change

May 2010

Article published in *Environment* magazine[93] as part of a collection of responses to the December 2009 climate negotiations at COP15 in Copenhagen

Marching up the hill ...

The rhetoric leading up to the Copenhagen Climate Summit last December (COP15) was deafening. Voices – some sombre, some shrill, some almost hysterical – told us that COP15 must deliver a deal 'to save the planet' and 'to protect civilisation as we know it.' These people characterised it as 'the last chance we have to tackle climate change.'[94] Such an atmosphere was not conducive to calm, considered, and realistic negotiating. And it was a task made harder because in recent years, so many other issues have been added to the tangled knot of climate change politics: the loss of biodiversity, the gross inequity in patterns of development, degradation of tropical forests, trade restrictions, violation of the rights of indigenous peoples, intellectual property rights, and others. The list seemed to grow by the month.

The world arrived at Copenhagen with a Rubik's cube climate-change puzzle containing just too many dimensions to be solved. Copenhagen was the culmination of 17 years of policy-making under the UN Framework Convention on Climate Change, of 12 years of prospecting about what was to come after the terminal 'Kyoto year' of 2012, and of two years of navigation using the Bali Road Map. COP15 attracted, ostensibly, the largest number of world leaders ever to attend a summit meeting. And the participation of civil society reached a new high for international climate negotiations, with an estimated 45,000 participants,

---

93 Hulme, M. (2010) Moving beyond climate change *Environment* 52(3), May/June, 15–19. A shortened version of this article also appeared on the *BBC News On-line Viewpoint* 11 May 2010. http://news.bbc.co.uk/1/hi/sci/tech/8673828.stm.

94 This phrase was used by UK Prime Minister Gordon Brown in a speech in London to the Major Economies Forum, 19 October 2009. His remarks were: 'If we do not reach a deal this time, let us be in no doubt; once the damage from unchecked emissions growth is done, no retrospective global agreement, in some future period, can undo that choice. By then it will be irretrievably too late.' Available at http://www.number10. gov.uk/Page21033.

their numbers swollen through eager anticipation of the role to be played by a new American president.

Underlying this political moment – this meeting that was claimed by some to be 'the most important in human history' – was a narrative of crisis. A climate crisis.[95] It was a narrative that drew its strength from various scientific assessments and reassessments of climate change, starting with the Intergovernmental Panel on Climate Change (IPCC) and including other statements from national scientific academies and expert groups.[96] This was the science, so it was claimed, which 'demanded' this and 'demanded' that from the world's political leaders. Or, as President Obama remarked, this was the science that 'dictates [that] even more needs to be done.'[97]

... and marching down again

But Copenhagen did not turn out as planned. No climate-saving deal was reached, not even a political agreement between the parties to the UN Framework Convention, let alone something that was legally binding. An accord was drafted – the Copenhagen Accord[98] – to which nations could opt in if they so chose. Some nations did; some did not. The Accord offered a nod to science. It 'recognised the scientific view that the increase in global temperature should be below 2°C.'[99] It also promised money – up to US$100 billion annually by the year 2020 for mitigation and adaptation activities in developing countries.

95 Numerous books about climate change are framed using the idea of 'crisis': D. Archer and S. Rahmstorf, *The Climate Crisis: An Introductory Guide to Climate Change* (Cambridge, UK: Cambridge University Press, (2010); A. Gore, *Our Choice: A Plan to Solve the Climate Crisis* (London, UK: Bloomsbury Publishing, 2009); M. Robinson, *America Debates Global Warming: Crisis or Myth?* (New York: Rosen Central, 2007); R. Gelbspan, *Boiling Point: How Politicians, Big Oil and Coal, Journalists and Activists Have Fuelled the Climate Crisis – and What We Can Do to Avert Disaster* (New York: Basic Books, 2004).

96 For example see the G8+5 Academies' joint statement released in July 2009: 'Climate Change and the Transformation of Energy Technologies for a Low Carbon Future'; and 'The Copenhagen Diagnosis: Updating the World on the Latest Climate Science,' a statement written by 26 leading scientific experts, November 2009. Available at http://www.ccrc.unsw.edu.au/Copenhagen/Copenhagen_Diagnosis_LOW.pdf.

97 This phrase is taken from comments made by President Obama at Copenhagen, 18 December 2009: 'We know [to-be-set emission targets] will not be by themselves sufficient to get to where we need to be. Science dictates even more needs to be done ... Ultimately this issue is going to be dictated by the science. The science indicates we are going to have to take more aggressive steps in future.' Available at http://www.whitehouse.gov/the-press-office/remarks-president-during-press-availability-copenhagen.

98 The full three-page Copenhagen Accord can be found here: http://unfccc.int/resource/docs/2009/cop15/eng/l07.pdf.

99 UNFCCC (2009) Copenhagen Accord p.1.

What was perhaps most striking about the dynamics of Copenhagen, however, was the unavoidable evidence of the shifting centres of geo-political power. No longer was it the European Union or the Anglophone nations that carried the day, nor even the nations of the Organisation for Economic Co-operation and Development. It was China, India, Brazil, and South Africa that became the makers and breakers of deals.

Copenhagen has shown us the limits of what can be achieved on climate change through centralisation and hyperbolic multi-lateralism. Climate change – least of all, the Rubik's cube version of climate change we have chosen to construct – will not be adequately defused through such top-down United Nations processes. Earth-system scientists may have shown us how the physical planetary system functions as a single entity, but we are a long way short of displaying even the minimum attributes necessary for effective earth system governance.[100]

Refocusing our goals

The current efforts to defuse the risks of climate change are therefore leading us nowhere. We have been dazzled by the sciences of the Earth system and have framed the problem as one of 'stabilising global climate'. We have been misled by the successes of the Montreal Protocol and have sought the solution through a single unified global framework. We have locked ourselves into seeking an unachievable goal – climate stabilisation – that is based on a scientific narrative that cannot bear the weight of expectation and that is to be delivered through the hubristic, centralised management of the planet and its inhabitants.

We must therefore now either redefine the goal or the delivery mechanism – or, even better, redefine both. Rather than be dictated to by this scientific narrative of avoiding climate chaos (and let us remind ourselves it is we who determine our goals, not some reified essentialist science), we need to rearticulate our goals, thereby allowing us to reframe the problem of climate change and to redesign our strategies.

I therefore suggest that our ultimate goal is not to 'stop climate change'. We have mistaken the means for the end. Our goal is surely to ensure that the basic human needs of the world's growing population are adequately met; that we move toward a development paradigm where we are living within our techno-ecological means and not beyond them; and that our societies are adequately equipped to withstand the risks and dangers that come from a changing climate. Distinguishing whether those risks and dangers are natural or not is hardly the point.

---

100 Frank Biermann wrote a paper in 2007, '"Earth system governance" as a cross-cutting theme of global change research,' *Global Environmental Change* 17, 3–4 (2007): 326–337, in which he called for earth system governance as a political programme of action. He outlined some of the characteristics of such a governance system.

Redesigning our strategies

These should be the goals of an aspirational, morally aware, and survival-oriented species. Refocusing our goals in this way – moving beyond climate change to see what lies on the other side – has implications for how we design our policy frameworks and implement our policy measures.

What might an alternative strategic approach to meeting these goals look like in practice? It should be an approach that allows us to take a few small steps that offer demonstrable, quick payback. It should be an approach that is polycentric in structure and pluralist in instinct. And it should be an approach that prizes substantial investment in technology innovation.[101]

The first step is to fragment the Rubik's cube. We need separate policy frameworks and interventions for short-lived and long-lived climate-forcing agents. The physical properties, sources, and policy levers of short-lived forcing agents – black soot, aerosols, methane, and tropospheric ozone – are quite different than are those for long-lived forcing agents – carbon dioxide, halocarbons, and nitrous oxide. Then we need to separate energy policy from climate policy. There is no obvious logic, for example, in connecting innovation in liquid fuel technologies with innovation in information and management systems that allow societies to manage climate risks. And third, we need to attend directly to the development demands from the world's bottom two billion. There is no essential reason why attending to the scandal of world poverty should be ransomed to reaching a global deal on designing a long-term, low-carbon energy revolution.

Below are six strategic policy initiatives that are not stitched together into one single impossible package – the progress of which is too easily hijacked by diversionary tactics such as arguing about whether or not the science behind a two-degree global temperature target – or indeed any global target – is sound.

Eradicate emissions of black carbon[102]

Black carbon warms the atmosphere at regional and global scales. It is also a public health hazard; 1.8 million people die every year from exposure to black carbon through indoor fires. It is feasible to nearly eradicate emissions of black carbon through targeted and enforced regulation. The equivalent of about 25 Gigatons of carbon would thus be removed from the atmosphere by 2050. The environmental payback here is relatively quick, with huge public health benefits.

---

101 In shaping the argument of this section, I am indebted to the ideas of Scott Barratt, Ted Nordhaus, Elinor Orstom, Roger Pielke Jr., Gwyn Prins and Michael Shellenberger.
102 See A.P. Grieshop, C.C.O. Reynolds, M. Kandlikar, and H. Dowlatabadi 'A Black-Carbon Mitigation Wedge,' *Nature Geosciences* 2 (2010): 533–534, for a discussion of why and how this can be achieved.

Work toward an integrated forest protocol[103]

Tropical forests are a key asset for humanity's future, not merely because of their carbon store, but also because of their husbandry of biodiversity, their timber and non-timber products, and their wider livelihood functions for indigenous peoples. Rather than seeking to lock tropical forest management into an all-embracing climate convention, and thus get snarled up in the complexities of reducing industrial carbon emissions, forests should be managed through a separate protocol that recognises the integrated value of these ecosystems. This needs to be a radically different version of the existing REDD (Reducing Emissions from Deforestation and Degradation) mechanism, which has an exclusive focus on reducing emissions of carbon dioxide by limiting deforestation and forest degradation.

Stimulate new climate risk management institutions

All societies are maladjusted to climate to some degree. In other words, climate extremes and variability impose costs on all societies (as well as generating benefits). It is therefore prudent to evolve technologies, institutions, and management practices that minimise the costs and damages wrought by climate, and even more so, to build this adaptive capacity as climates – and consequential risks – change. The new initiative from the World Meteorological Organisation for a Global Framework for Climate Services is one example of new climate risk management strategies. This proposed new Global Framework seeks to strengthen, worldwide, the production, availability, delivery, and application of science-based climate prediction and associated services, so as to improve the capacity of societies to manage climate risk. Others include new micro-insurance initiatives, community-based adaptation, and improved early warning systems. These initiatives and the sharing of good adaptation practice make sense, whether or not climate risks are being changed by human activities or how quickly they are changing.

Honour UN General Assembly Resolution 2626, 24 October 1970

Without negotiating new treaties, without setting up new funds and new institutions to manage new funds, and without arguing over how much and from whom, there is a simple step that can be taken to improve the life chances for many. A simple honouring of the resolve made in 1970 by the OECD nations to

---

103 See papers by O. Venter, W.F. Laurance, T. Iawamura, K.A. Wilson, R.A. Fuller, and H.P. Possingham 'Harnessing Carbon Payments to Protect Biodiversity,' *Science* 326 (2009): 1368; and U.M. Persson and C. Azar, 'Preserving the World's Tropical Forests – a Price on Carbon May Not Do,' *Environmental Science & Technology* 44 (2010): 210–215, for discussion of moving beyond the REDD mechanism.

commit 0.7 per cent of Gross National Product to overseas development assistance would, at one stroke, release between $200 and $250 billion annually into meeting the development needs of the world's bottom two billion. It was a resolve renewed by the OECD nations at Monterrey in 2002: 'to make concrete efforts towards the target of 0.7 per cent of gross national product as overseas development assistance to developing countries.' Currently, not much more than half of this promised amount ends up as development assistance.

This investment in human welfare compares favourably with the US$100 billion promised under the Copenhagen Accord for 2020. It is not about defusing the effects of climate change on future generations. This is an issue of social justice and humanitarian welfare, here and now. It would not solve climate change, nor would it eradicate poverty. But it would be a gesture a hundred times more important for the future of the world than the vague promises for 2020 that emerged from Copenhagen.

Use the Montreal Protocol[104]

Between 5 and 10 per cent of human forcing of the climate system originates from a very specific group of long-lived gases: for example, the ozone-depleting and non-ozone-depleting halocarbons and sulfur hexafluoride. These chemicals are used in the refrigeration and electrical and other industries. The sub-set of these gases which are climate-warming but non-ozone-depleting, could be controlled and then eliminated through a modification of the Montreal Protocol. This protocol has been successful for controlling ozone-depleting substances by tackling production rather than consumption, and it could do likewise for similar classes of industrial gases. There is no need to tie this initiative to the much more challenging task of reducing fossil carbon emissions.

Long-term decarbonisation

Emissions of carbon dioxide from energy production comprise no more than 50 per cent of the human forcing of the climate system. These emissions are the hardest to reduce, and reductions bring benefits only in the longer term; yet we have been obsessed with this particular cause of climate change. Somewhat perversely, we have framed the problem of climate change and designed our policy frameworks to tackle the harder half of the problem and not the easier half! There are many good reasons for accelerating the decarbonising of our energy supply, but to do so will require substantial technological innovation.

---

104 See the papers by G.J.M. Velders, S.O. Anderson, J.S. Daniel, D.W. Fahey, and M. McFarland, 'The Importance of the Montreal Protocol in Protecting Climate,' *Proceedings of the National Academy of Sciences* 104, no. 12 (2007): 4814–4819; and J. Cohen, A. Rau, and K. Bruning, 'Bridging the Montreal–Kyoto Gap,' *Science* 326 (2009): 940–941.

Carbon-trading mechanisms will not be sufficient; the emergent carbon markets of Europe and the Clean Development Mechanism have been abject failures in constraining carbon emissions. What are needed here are commitments to innovation policy and large investments in new energy technologies. These can be achieved through unilateral, bilateral, or minilateral initiatives; it does not require a global treaty signed by 193 countries. One way to secure the needed investment is through a hypothecated carbon tax, initially low and rising gradually.[105] This is a long-term transition, which can afford to build slowly.

Technologies of humility

We will only clearly see how mesmerised we have become by the idea of anthropogenic climate change when the history of the period 1985 to 2010 is written. We will then see how the knowledge community pushed the claims of climate prediction, thereby lending credence to the idea that global climate stabilisation was possible. We will see how, as the new world order of the 1990s emerged, we placed unwarranted faith in centralised global environmental governance through the United Nations. We will see how we fixated on tackling fossil carbon–forcing of climate to the detriment of tackling the easier parts of the problem. And we will hopefully then see the significance of the Copenhagen meeting in December 2009, a meeting after which we began to recast the problem of climate change and to redesign our strategies to move beyond it.

The world has no meta-narrative – whether offered by science, ideology, or religion – that can provide a universally accepted foundation upon which a centralised global governance regime can be built. The claims that Earth system science might provide such a foundation have weakened rather than strengthened. The IPCC has discovered the limit of its influence. We are beginning to recognise climate change for the wicked problem that it is, and we will now embark on a more diverse, less authoritarian search for clumsy solutions.[106]

In going beyond climate change – lowering our ambitions and seeking small steps in a polycentric world of pluralist views and preferences – we paradoxically increase the likelihood that we can deliver on at least some of our aspirations for a better world. It is not more certain scientific predictions that we need, nor a charismatic leader to arise from 'the east', nor grand dreams of creating a global thermostat in the sky above. It is what Sheila Jasanoff, Professor of Science and Technology Studies at the John F. Kennedy School of Government at Harvard

---

105 See I. Galiana and C. Green, 'Let the Global Technology Race Begin,' *Nature* 462 (2009): 570–571, for a discussion about why this is necessary and how it may be facilitated.
106 These are ideas that I explore in some detail in my book *Why We Disagree About Climate Change* (Cambridge, UK: Cambridge University Press, 2009).

University, has referred to as the 'technologies of humility'[107] – 'disciplined methods to accommodate the partiality of scientific knowledge and to act under irredeemable uncertainty' – that will offer us the best prospects for taming the risks of climate change.

107 Jasanoff, S. (2007) Technologies of Humility *Nature* 450, 33 (1 November 2007).

35 Climate intervention schemes could be undone by geopolitics

7 June 2010

Essay written for *Yale Environment 360,* on-line magazine[108]

As global warming intensifies, demands for human manipulation of the climate system are likely to grow. But carrying out geoengineering plans could prove daunting, as conflicts erupt over the unintended regional consequences of climate intervention and over who is entitled to deploy climate-altering technologies.

Last month, J. Craig Venter announced that his team had successfully developed the first self-replicating cell to be controlled entirely by synthetic DNA. Not artificial life exactly, but certainly something different: a synthetic cell in which humans had intervened deliberately with the express purpose of changing the genetic structure and characteristics of a natural organism.

Humans are lining up comparable purposeful interventions in the functioning of another physical system – not the microscopic system of a bacterium, but the macroscopic planetary system that fashions and delivers all our climates. The range of such potential climate intervention technologies – from altering how much of the sun's energy strikes the Earth, to removing carbon dioxide from the atmosphere – continues to expand against a backdrop of anxiety that humanity may inadvertently be pushing global climate toward a dangerous state.

These two new ventures – manipulating the biological functions of cells and manipulating the physical functioning of the climate system – may be seen as simply the latest steps in the enduring human project of seeking control over the physical world. Hominid mastery of fire in the Paleolithic brought about radical changes in the possibilities for human life, and the manufacture of antibiotic drugs in the 20th century opened up a wide range of new medical treatments that have reduced suffering and extended human life. Designing self-replicating cells

108 Hulme, M. (2010) Climate intervention schemes could be undone by geopolitics *Yale Environment 360* 7 June http://e360.yale.edu/content/feature.msp?id=2283.

and re-tuning global climate may therefore appear as inevitable developments in our ingenuity and our ability to manipulate the world around us.

But compared to the questions raised by Venter's biotechnologies, two categorically different sets of questions arise about climate manipulation: How do we judge the risks of unintended consequences? And who is entitled to initiate the large-scale deployment of a climate intervention technology – and under what circumstances?

Proponents are suggesting two broad categories of technologies to roll back global warming. The first, solar radiation management, calls for altering the solar radiation budget of the planet, using such technologies as mirrors in space, aerosols in the stratosphere, and cloud whitening over the oceans. And then there are technologies, grouped under the category of carbon dioxide removal, that propose to accelerate the removal of carbon dioxide from the atmosphere by fertilising the oceans with iron, extracting carbon dioxide from the atmosphere, or sequestering carbon dioxide by heating biomass in oxygen-free kilns and burying the charcoal underground.

Such interventions would bring about, if not exactly artificial climates, then certainly synthetic ones. The calls for significant investments in these technologies have grown in boldness and urgency over the last few years. Whether from government agencies or private investors such as Richard Branson or the company, Climos, resources are being directed into pursuing something akin to Venter's vision of synthetically controlled cells, but the 'cell' in question here is the planetary climate.

Both genres of climate intervention technologies raise serious ethical questions about the propriety of such manipulations, about their accordance with the collective will of people on Earth, and about the unforeseen side effects of such interventions. But the proposition of creating synthetic climates through solar radiation management (less so with carbon dioxide removal) introduces a range of additional concerns not shared with microscopic cellular manipulation. These concerns arise from the brute fact that there is only one climate system with which to experiment, and it is the one we live with. If it is planetary-scale manipulation of climate that is desired – and it is – then experimentation has to be conducted on a planetary scale to prove the effectiveness – or not – of the technology.

The first concern is the risk of unintended consequences. Given that it is not possible to conduct large-scale planetary experiments in solar radiation management before going 'live' with the technology, risk assessments have to fall back on using virtual climates generated by computer models. The Earth system models currently used to explore the possible future effects of rising atmospheric concentrations of greenhouse gases are the same ones that have to be used to explore the simulated consequences of a variety of solar radiation interventions.

Using aerosols to offset the additional planetary heating caused by greenhouse gases is a relatively straightforward theoretical calculation; it is a case of simple planetary budgeting. Much harder is to know what this 're-balancing' of the global heat budget will do to atmospheric and ocean dynamics around the world. These are the dynamics that make weather happen at particular times and in particular places and which – through various combinations of rain, wind, temperature, and humidity – shape ecological processes and human social practices. The dangers and opportunities associated with climate occur through these local weather phenomena, not through an abstract index of global temperature.

If the goal of climate engineering is simply to reset the global temperature dial at its nineteenth or late-twentieth century register, that might be possible to do. But in the process of doing so, significant perturbations to regional climate conditions, and inter-annual variability around those conditions, are likely to be introduced. Even if changes in the frequency and intensity of storms and precipitation were to be a zero-sum game globally, the distributional effects of such changes will create winners and losers. Such phenomena as El Niño, the Asian monsoon, and the Arctic Oscillation will not remain unaffected. And given the far-from-adequate ability of Earth system models to simulate the regional-scale dynamics of the hydrological system, no one should be confident that the full risks of solar radiation management interventions will be revealed and quantified.

Which brings us to the second question that sets apart the project to fashion a synthetic climate from the project to create synthetic self-replicating cells: Under what future scenario could one imagine full-scale deployment of solar radiation management taking place? Many commentators have drawn attention to the multi-layered issues of financing, ethics, governance, geopolitics, and public opinion that surround most of these solar radiation intervention technologies. These were very much to the fore at the recent Asilomar International Conference on Climate Intervention Technologies in California earlier this year.

And yet a number of senior and significant voices in the scientific academy and policy community continue to speak of the urgency with which solar radiation management research should be pursued. They offer these putative control technologies as another option in the portfolio of climate management strategies, with climate manipulation joining climate change mitigation and climate adaptation in a trinity of strategies available for policymakers. At the very least, it is argued, solar radiation management should be available as a backstop technology if the world finds itself in a climate emergency when a dangerous tipping point needs to be avoided.

But can we imagine a possible scenario under which the decision to proceed to full deployment of solar radiation management might be made? Let us assume the injection of aerosols into the stratosphere had been placed at the top of the list of climate intervention technologies. Let us also assume that the basic operational mechanics of getting aerosols into the optimal layers of the stratosphere for maximum solar shielding had been figured out. One possible scenario might look something like this:

It is January 2028 and the UK – one of the permanent members of the UN Security Council – puts forward a formal resolution to start the systematic injection of sulphate aerosols into the stratosphere. The UK's argument is that with Arctic sea ice extent the previous summer having shrunk to just 25 per cent of its late-twentieth century value, with monitors in Canadian permafrost identifying increased rates of methane release, and with the explosion at a nuclear reactor in China two years earlier leading to a moratorium on all new nuclear power plant construction, such direct climate remediation measures are called for.

The Intergovernmental Panel on Climate Change (IPCC) provides a report for the Security Council on the regional climatic risks of such intervention. Based on the best Earth system models, the IPCC offers probabilistic predictions of the 10-year mean changes in regional rainfall around the world that would result from sustained aerosol injection.

The 15 members of the Security Council argue over the evidence. In particular, they spend much time weighing the probabilities that the Asian monsoon might be weakened as a result. Security Council members also argue about how long the initial aerosol injection should continue – for 1 year, 3 years, or 5 years. Against a background of vociferous, and at times violent, globally-coordinated public campaigns (both in favour and against such intervention), the Security Council votes 11–2 in favour, with 2 abstentions. The deployment will proceed for a one-year period, after which a full evaluation will be conducted.

Over the following months, protestors attempt to sabotage some of the planes being used to inject aerosols, and direct-action groups affiliated with HOME (Hands Off Mother Earth) send up their own aircraft in symbolic efforts to scrub the aerosols from the stratosphere. After one year the deployment is temporarily halted and climate data are evaluated.

Global temperature has indeed fallen from the previous 10-year mean of 15.23°C (the 1961–1990 average was 14°C) to just 14.57°C, the coolest year on the planet since 2014. But regional climate anomalies have been large and variable. Of most concern was a failure of the Asian monsoon, at the cost of $50 billion to the Indian economy, and the most intense cyclone season in the South China Sea for 20 years.

India – one of the rotating members of the Security Council – and China now trigger an emergency debate calling for a permanent ban on deployment of aerosol injection technologies. The IPCC argues that one year's data prove nothing about the efficacy or impact of solar radiation management. But against a background of further global protests, led by the new popular civic movements in China and India, the Security Council now splits 5–5, with 5 abstentions. Turmoil ensues as two Canadian billionaires unilaterally continue aerosol injection.

Of course one could create a hundred other scenarios under which the story of solar radiation management may unfold. But I use this one to draw attention to the profound political obstacles and humanitarian risks that shadow attempts to engineer the climate through solar radiation management. The organisation HOME already exists, seeking to mobilise people everywhere to tell climate engineers to proceed no further with climate manipulation.

The technical body supporting the work of the UN Convention on Biological Diversity has recently proposed a draft text[109] along the following lines: 'No climate-related geo-engineering activities [should] take place until there is an *adequate* scientific basis on which to justify such activities and *appropriate* consideration of the associated risks for the environment and biodiversity and associated social, economic and cultural impacts' (emphasis added).

Words such as 'adequate' and 'appropriate' offer new grounds for contention in an already argumentative world. If the politics of climate mitigation policy under the guise of the Kyoto Protocol have proved intractable, just wait until we see the geopolitics surrounding the negotiation of the first protocol on engineering synthetic climates. In the name of saving the planet from inadvertent greenhouse-gas exacerbated climate change, climate engineers may simply be offering us one Promethean fire to offset the effects of another.

---

109 CBD (2010) COP10 (Nagoya) Decision X/33 Biodiversity and climate change Section 8/w.

36 On the 'two degrees' climate policy target

June 2012

Contribution to the book *'Climate change, justice and sustainability: linking climate and development policy'* edited by Ottmar Edenhofer and colleagues[110]

The formal adoption of a global temperature target to drive, or at least guide, climate policy development dates back to the mid-1990s. The origins and history of the 'two degrees' target and how it was adopted by the EU Council in 1996 – and re-affirmed in 2007 – has been well told in articles by Tol (2007) and Randalls (2010). In recent years the two degrees target has gained in visibility, both in public discourse and in policy deliberations. For example, it underpins the UK's 2008 Climate Change Act and was commended by the G8 meeting at L'Aquila in July 2009. It was also given prominence in the Copenhagen Accord which emerged from the UNFCCC meeting in December 2009 in Copenhagen.

'Two degrees' – limiting the rise in globally-averaged temperature to no more than 2°C above the pre-industrial level – has become the benchmark for policy advocacy around climate change and for many environmental and climate justice campaigns. It has also functioned as an anchoring device (Van der Sluijs *et al.*, 1998) in climate science-policy analysis and interaction. And this single index of climate performance – collapsing the complexity and diversity of weather and climate around the world into global temperature – has gained powerful iconic and cultural status. But does the world need a global target in order to drive and guide climate policy? And if it does, is global temperature the most suitable index to use? In this short essay I identify four characteristics of the 'two degrees' target, namely: universality; ambiguity; doubtful achievability; and questionable legitimacy. I suggest why each of these characteristics undermines the value and necessity of such a target. I conclude the essay by contrasting the 'two degrees' target with other types of targets developed and used in other areas of public policy-making.

110 Hulme, M. (2012) On the 'two degrees' climate policy target pp.122–125 in: *Climate change, justice and sustainability: linking climate and development policy* Edenhofer, O., Wallacher, J., Lotze-Campen, H., Reder, M., Knopf, B, and Müller, J. (eds), Springer, Dordrecht/Heidelberg, Germany. Reproduced with kind permission from Springer Science+Business Media B.V.

The 'two degrees' target is, by definition, *universal*. It offers one numerical index by which to judge the future behaviour of the global climate system. It suggests that climate policy effectiveness should ultimately only be judged against whether it contributes to achieving this one universal goal. It draws attention away from the desirability of a wider set of more diverse climate policy goals which may have greater regional or national legitimacy and traction and which may be easier to implement. Ostrom's (2010) proposal to approach 'global commons' problems through polycentric policy initiatives may be impaired by the imposition of a 'one-size-fits-all' approach to policy orientation.

The 'two degrees' target is also highly abstract. Global temperature has no resonance with the everyday experience of weather and climate. It is a constructed quantity. Whether cognitively or existentially it has difficulty engaging the human imagination. Thinking of environmental and social policy through the lens of 'two degrees' also opens the way for the emergent discourse around climate engineering: deliberate manipulation of the planetary atmosphere and oceans to achieve an outcome measured in terms of this one index – global temperature. The metaphor of a global thermostat is a powerful one, but it opens up new frontiers for geopolitical tensions about what the thermostat setting should be and who controls it. Whether viewed rhetorically or pragmatically, adopting such a universal target to guide the conduct of affairs between nations is dangerous and of limited value.

The second characteristic of the 'two degrees' target is *its ambiguity*. Global temperature is an 'output index' of the climate system rather than corresponding to the range of underlying human 'input factors' (such as greenhouse gas and aerosol emissions, land cover, population). And because the relationship between input (human forcing) and output (global temperature) is deeply uncertain, agreeing a target of 'two degrees' helps little in specifying what the various input factors should be. Owing to the wide range of possible values of the climate sensitivity and the deep uncertainty about the aggregate global effects of forcing agents such as aerosols and black carbon (notwithstanding the extent of natural climate variability – see below), the 'two degrees' target is compatible with a very wide range of input scenarios. If it *were* deemed global target setting was necessary for climate policy, then a carbon dioxide concentration target, for example, would be much less ambiguous. This has indeed been recognised by many campaigners, such as Bill McKibbin's social movement '350.org' which campaigns for an atmospheric carbon dioxide concentration of 350ppm.

Third, the 'two degrees' target *is unattainable*. I don't mean here in the sense that politics, economics, culture and technology may conspire to prevent the necessary emissions reductions (though they may). I mean in a wider sense that manoeuvring the world's development pathway to deliver a re-stabilised global climate at no more than 2°C above pre-industrial levels requires a higher level of understanding than we have of the climate system and the extent of human influence upon it. It implies that the only factors affecting global temperature are human factors and ignores the extent of natural climate variability. Global temperature varies on multiple time-scales for complex reasons and current understanding suggests that on multi-decadal timescales natural forcing of the climate system could account for up to 0.5°C warming or cooling. This is 25 per cent of the target temperature rise and adds further ambiguity about its achievability (see above). Believing that a 'two degree' world can be engineered suggests a level of managerial control of the planetary system that humans are never likely to attain.

Finally, I consider the 'two degrees' target in terms of its *legitimacy*. Who has established this goal of international climate policy and who has the right to establish it? What is interesting here is that neither scientists nor politicians are willing to fully accept responsibility for its adoption. The scientific community – as given voice through the IPCC – assiduously makes clear that identifying a target for climate policy is a value-laden judgement and therefore falls beyond the remit of scientific enquiry. Yet the political community – whether advocacy campaigns or national and international politicians – continue to defer to 'what the science demands'. The Copenhagen Accord, for example, recognised 'the scientific [sic] view that the increase in global temperature should be below 2 degrees Celsius' (UNFCCC, 2009: 1).

I believe these four characteristics of the 'two degrees' target – its universality, its ambiguity, its unachievability and its illegitimacy – challenge its validity and necessity for climate policy-making. Counter-arguments in *favour* of this target include that: (a) it 'is demanded by the science' and (b) it usefully focuses the political mind. But as we have seen above, 'two degrees' is not *demanded* by science any more than science *demands* a target of zero degrees or of four degrees. There is not a global temperature target waiting to be *discovered* by scientific enquiry. And it is debatable whether the 'two degrees' target in fact inhibits policy-making at an international level. As discussed above, global temperature is an abstract index of planetary behaviour which hides deep ambiguity in the relationship between input control factors and output performance. The 'two degree' target – being abstract, distant in time and ambiguous – is as likely to allow politicians to evade its demands as to encourage them to embrace them.

It can also be argued that the 'two degrees' target is socially regressive, or at best diversionary. It runs the danger of confusing ends with means. It is not a global climate system delivering some abstract global temperature – whether zero, two or four degrees above the nineteenth century level – that is a public good. The ultimate goals of progressive environmental and social welfare policy revolve around individual and collective human well-being (unless one adopts a strong non-anthropocentric ethic). Thus the end goal of such policy must surely include reducing global poverty, improving literacy and educational opportunity, empowering citizens, etc. Elevating 'two degrees' to the ultimate goal of climate policy development may endanger this human welfare agenda.

The 'two degrees' target sits easily within a managerial audit culture which has come to dominate (at least) European societies in recent years, with numerical public sector league tables and performance targets. The danger of such highly quantified audit cultures is that one may hit the narrowly defined numerical target, but miss the desired underlying welfare goals of the policy intervention.

With climate change the example of this would be a policy intervention that secured the 'two degrees' target through intrusive large-scale biogeophysical engineering of the atmosphere or oceans, but which did nothing to, say, alleviate poverty, achieve universal energy access or improve female literacy.

In conclusion I wish to contrast the case of climate change and 'two degrees' with other areas of public policy where different types and levels of target-setting are introduced. Take the example of public health. National health ministries do not rhetoricise or plan around a national goal to increase average life expectancy by 'x' years within 'y' decades. If they did set such targets they would be seen as purely aspirational, with little tangible value for health policy-making. The relationships between the range of input (health risk) factors and output performance (average life expectancy across a population) are too complex and unknown. Instead, national health policy targets are much more narrowly prescribed – for example, different treatment rates for different forms of cancer, screening and vaccination programmes, dietary guidelines. Such a fragmented approach to public health policy facilitates more pragmatic, targeted, accountable – and hence achievable – management interventions. A *by-product* of their implementation will be to increase life expectancy.

An example closer to the climate change case would be the Millennium Development Goals (MDGs). A key difference here is that specification of the MDGs is explicitly political – it is not claimed that they derive from a scientific analysis which warrants one set of targets over any other ones. The MDGs are focused on very specific welfare goals, unlike the 'two degrees' target which is several (ambiguous) steps removed from delivering tangible welfare gains on the ground.

A final example would be to ask why the world has been willing to embrace a universal global temperature target, but has kept well away from the adoption of a global population target. A global population target would be heavily contested for all of the reasons I have suggested above that afflict the 'two degree' target: universality, ambiguity, unachievability and illegitimacy. And yet these reasons have not prevented 'two degrees' emerging as the goal around which climate policy rhetorically congregates. I suspect one of the reasons for this has to do with the different ways in which scientific knowledge claims – deriving from Earth system modelling in the case of climate change – have interacted with political and ethical argumentation. Science has been used to trump political, ethical and religious argumentation in the case of climate change, but not in the case of population policy – where a global population target is recognised as being undesirable and infeasible.

Section six

Communicating

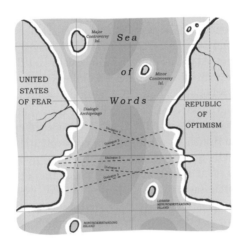

Introduction

As a matter of public and policy concern, climate change reached its greatest saliency in the autumn of 2006, certainly in the UK, maybe in the United States and perhaps elsewhere. Al Gore's film *An Inconvenient Truth* had appeared in May that year, the UK Government's Stern Review of the economics of climate change was published in October and a new UK coalition of campaigning NGOs – *Stop Climate Chaos* – had formed to begin lobbying for a Parliamentary Climate Change Bill. Results from the new IPCC 4th Assessment Report were beginning to be trailed and the British Government was preparing to bring climate change under debate at the UN Security Council.

It was at this moment that I wrote the first of the articles in this section (Chapter 37), a short essay which was posted on *BBC News On-Line* in early November 2006, having previously been rejected by a couple of national newspapers. The thrust of the essay was concern about the language being used to communicate the risks of climate change – 'chaos', 'catastrophe', 'weapon of mass destruction' – language which, I argued, over-extended scientific evidence and which, even in campaigning terms, would be self-defeating. 'The chaotic world of climate truth' was written during my final year as Director of the Tyndall Centre. This essay was my first public articulation of my growing concern over a number of years about the ways in which climate science was being deployed in public discourse.

The article generated a large volume of correspondence from scientists, members of the public and campaigners alike (getting a reaction is usually a sign of a worthwhile intervention). Many of these correspondents thanked me for challenging the over-heated alarmist rhetoric, as in this member of the public who thanked me for 'your call for sanity in the climate change debate, which seems to be completely in the grip of Private Frazer Syndrome ('We're all doomed!').' The article also triggered a full-blown response in the peer-reviewed academic literature in which Australian climate scientist James Risbey challenged me for failing to distinguish between being 'alarmist' and communicating findings which were 'alarming' (Risbey, 2008).

The second item in this section (Chapter 38) is a review for the magazine of the British higher education system of a trade book from Bloomsbury Press – *The Hot Topic: how to tackle global warming and still keep the lights on* – written by Sir

David King with science journalist Gabrielle Walker. The book was published early in 2008 to mark the drawing to a close of the former's eight-year appointment as the UK Government's Chief Scientific Advisor. I reproduce the essay here because it explains why I am critical of a particular way of framing climate change, namely as a matter of science, technology and economics. This frame was dominant throughout King's book, which therefore fails to engage with the social and cultural dynamics by which climate change gains its public meaning and cultural significance. In this sense *The Hot Topic* is a poor example of communicating what climate change is really about.

Another example of inadequate communication is the focus of my next selection (Chapter 39), an essay written for the on-line American *SeedMagazine*. This was written on my way back from attending the Copenhagen Climate Change Congress which took place in March 2009 and was the (then) largest single gathering of climate change researchers. I was interested in the Six Key Messages which on the final day of the Congress were reported around the world as a collective statement from scientists to politicians ahead of COP15 to be held later that year, also at Copenhagen. In my essay I question whose views the Key Messages represented and what were 'the actions' being called for. The processes of scientific and scholarly deliberation and the roles of elite policy advocacy got badly confused in this exercise.

A couple of months after the Copenhagen Congress my book *Why We Disagree About Climate Change* (see Section Nine) was published. I sought to distil and distribute some of the arguments from this book in a number of different outlets, including the next selected item (Chapter 40) which was an essay published in September 2009 in the London-based *New Scientist* magazine. This essay summarises my argument about the different enduring myths through which the idea of climate change can be understood, myths that capture the human instincts of fear, hubris, nostalgia and justice. This construction is linked to the concept of framing, which begins to explain how it is that such a wide range of human responses can be invoked when presented with ostensibly the same factual information.

By early 2010, the discursive 'climate' of the climate change debate had significantly altered, certainly in Anglophone countries (see Chapter 50) and there was a more explicit and emboldened challenge being offered to the scientific and policy orthodoxy by a variety of commentators. I was invited by the editor of the in-house magazine of the UK's Royal Society of Arts, Manufactures and Commerce to write 'a feature for our forthcoming issue about why opinions are still so divergent about climate change and what we can do to persuade sceptics of the scale of the problem.' I did so and this essay is reproduced here as Chapter 41. I explain why opinions 'are still so divergent', but also point out that it matters who is the 'we' in the desire to persuade opponents. There are more than two sides to the questions raised by climate change and many of the different positions adopted involve complex combinations of beliefs, values and different attitudes to risk, technology and democracy.

The last of the items selected for this section (Chapter 42) was first published in Australia and is a contribution I made to a series of commentaries on the on-line ideas forum *The Conversation* called 'Clearing up the climate debate'. This series was posted in the middle (June–July 2011) of a vigorous public argument in Australia about new legislation for a carbon tax being pushed through Parliament by the Prime Minister Julia Gillard. I designed my intervention to draw out the significance for public debate of the multiple ways in which climate change gets framed – there can be no single clear story of climate change, only a selection of stories each of which emphasises (and therefore de-emphasises) one or more elements of the phenomenon more than others. It is important for scientists, citizens, campaigners and politicians to be alert to the different framing possibilities and strategies being deployed by different actors in public settings.

37 Chaotic world of climate truth

4 November 2006

An essay written for BBC News On-line Viewpoint[111]

As activists organised by the group Stop Climate Chaos gather in London to demand action, one of Britain's top climate scientists says the language of chaos and catastrophe has got out of hand

Climate change is a reality, and science confirms that human activities are heavily implicated in this change. But over the last few years a new environmental phenomenon has been constructed in this country – the phenomenon of 'catastrophic' climate change. It seems that mere 'climate change' was not going to be bad enough, and so now it must be 'catastrophic' to be worthy of attention.

The increasing use of this pejorative term – and its bedfellow qualifiers 'chaotic', 'irreversible', 'rapid' – has altered the public discourse around climate change. This discourse is now characterised by phrases such as 'climate change is worse than we thought', that we are approaching 'irreversible tipping in the Earth's climate', and that we are 'at the point of no return'.

I have found myself increasingly chastised by climate change campaigners when my public statements and lectures on climate change have not satisfied their thirst for environmental drama and exaggerated rhetoric. It seems that it is we, the professional climate scientists, who are now the (catastrophe) sceptics. How the wheel turns.

Boarding the bandwagon

Some recent examples of the catastrophists include Tony Blair, who a few weeks back warned in an open letter to EU head of states: 'We have a window of only 10–15 years to take the steps we need to avoid crossing a catastrophic tipping point.' Today, a mass demonstration in Trafalgar Square will protest, aiming to

111 Hulme, M. (2006) Chaotic world of climate truth *BBC News On-line, Viewpoint*, posted 4 November 2006 news.bbc.co.uk/1/hi/sci/tech/6115644.stm.

'stop climate chaos' – the name for a coalition of environmental activists and faith-based organisations. The BBC broadcast in May its Climate Chaos season of programmes. There is even a publicly funded science research project called Rapid.

Why is it not just campaigners, but politicians and scientists too, who are openly confusing the language of fear, terror and disaster with the observable physical reality of climate change, actively ignoring the careful hedging which surrounds science's predictions? James Lovelock's book *The Revenge of Gaia* takes this discourse to its logical endpoint – the end of human civilisation itself. What has pushed the debate between climate change scientists and climate sceptics to now being between climate change scientists and climate alarmists?

I believe there are three factors now at work. First, the discourse of catastrophe is a campaigning device being mobilised in the context of failing UK and Kyoto Protocol targets to reduce emissions of carbon dioxide. The signatories to this UN protocol will not deliver on their obligations. This bursting of the campaigning bubble requires a determined reaction to raise the stakes – the language of climate catastrophe nicely fits the bill. Hence we now have the militancy of the Stop Climate Chaos activists and the megaphone journalism of *The Independent* newspaper, with supporting rhetoric from the prime minister and senior government scientists. Others suggest that the sleeping giants of the Gaian Earth system are being roused from their millennia of slumber to wreck havoc on humanity.

Second, the discourse of catastrophe is a political and rhetorical device to change the frame of reference for the emerging negotiations around what happens when the Kyoto Protocol runs out after 2012. The Exeter conference of February 2005 on 'Avoiding dangerous climate change' served the government's purposes of softening-up the G8 Gleneagles summit through a frenzied week of 'climate change is worse than we thought' news reporting and group-think. By stage-managing the new language of catastrophe, the conference itself became a tipping point in the way that climate change is discussed in public.

Third, the discourse of catastrophe allows some space for the retrenchment of science budgets. It is a short step from claiming these catastrophic risks have physical reality, saliency and are imminent, to implying that one more 'big push' of funding will allow science to quantify them objectively.

We need to take a deep breath and pause.

Fear and terror

The language of catastrophe is not the language of science. It will not be visible in next year's global assessment from the world authority of the Intergovernmental Panel on Climate Change (IPCC). To state that climate change will be 'catastrophic' hides a cascade of value-laden assumptions which do not emerge from empirical or theoretical science. Is any amount of climate change catastrophic? Catastrophic for whom, for where, and by when? What index is being used to measure the catastrophe?

The language of fear and terror operates as an ever-weakening vehicle for effective communication or inducement for behavioural change. This has been seen in other areas of public health risk. Empirical work in relation to climate change communication and public perception shows that it operates here too. Framing climate change as an issue which evokes fear and personal stress becomes a self-fulfilling prophecy. By 'sexing it up' we exacerbate, through psychological amplifiers, the very risks we are trying to ward off.

The careless (or conspiratorial?) translation of concern about Saddam Hussein's putative military threat into the case for Weapons of Mass Destruction (WMD) has had major geopolitical repercussions. We need to make sure the agents and agencies in our society which would seek to amplify climate change risks do not lead us down a similar counter-productive pathway. The IPCC scenarios of future climate change – warming somewhere between 1.4 and 5.8 degrees Celsius by 2100 – are significant enough without invoking catastrophe and chaos as unguided weapons with which forlornly to threaten society into behavioural change.

I believe climate change is real, must be faced and action taken. But the discourse of catastrophe is in danger of tipping society onto a negative, depressive and reactionary trajectory.

38 Less heat, more light, please

21 February 2008

A review of: *'The Hot Topic: How to Tackle Global Warming and Still Keep the Lights on'* by Gabrielle Walker and Sir David King, written for *The Times Higher Education*[112]

Tackling climate change is about a lot more than understanding the science, insists Mike Hulme

Is climate change an idea that we invent and talk about or is it a physical phenomenon that we observe and quantify? It is of course both: an idea that exists in our social discourse and a phenomenon that exists in an external physical reality. In this hybrid character of climate change lies its fascination for us and also its intractability.

The Greeks were the first people we know to have conceptualised climate (different climates for them were a function of latitude) and to discuss climate change (they observed that changing the use of land could modify local climates). In the subsequent 2,500 years, our ideas about the changeability of climate and our abilities to observe and quantify that changeability have undergone several distinct phases.

We have arrived in the twenty-first century with a number of new tools with which to examine our changing climate and with a number of new stories about what it means for us. There is little doubt that the collective impact on the atmosphere of 6.6 billion living people (plus an equal number of dead ones) is altering the ways in which the Earth's physical system is working. We can theorise this impact, we can observe its consequences and we can make tentative predictions decades, even centuries, into the future about what this transformation may presage.

112 Hulme, M. (2008) Less heat, more light, please: Book review of: 'The hot topic: how to tackle global warming and still keep the lights on' by Walker, G. and King, D. in *Times Higher Education* 21 February, p.50. A version of this review also appeared in *The Independent* newspaper on 18 January 2009.

Equally, the idea of climate change, and how it relates to our wider human instincts for preservation, justice, comfort and power, has taken on a different character. Thus we 'fear' climate change and seek to 'control' and 'engineer' climate to some human-defined state of stability.

Which brings us to Gabrielle Walker and Sir David King's new book on the subject of climate change: *The Hot Topic*. Their credentials as scientists are impeccable and complementary: Walker, an award-winning science journalist and King, the Government's Chief Scientific Adviser from 2001 to 2008. This neat combination of experience offers something new amid the recent avalanche of books about climate change: something that is more considered than George Monbiot's *Heat*, more synoptic than Tim Flannery's *The Weather Makers* and less polemical than James Lovelock's *The Revenge of Gaia*.

But in the end, *The Hot Topic* remains a book about climate change written by two scientists. And herein lies its main weakness. For a scientist to write a book about climate change would be like a Catholic theologian writing a book about the nature of human consciousness: we definitely need this perspective, but it's hardly the whole story. Walker and King offer us a science, engineering and technology reading of climate change, with some economics and politics thrown in. The offerings from the social sciences, the arts and humanities, from religion and ethics are meagre indeed. Culture is mentioned three times in the book, public opinion and ethics just once each, the latter condescendingly referred to as 'touchy-feely choices' and set against the authoritative voice of science.

This unbalanced set of analytical tools that Walker and King bring to the issue of climate change reveals both the power and the weakness of the current framing of the phenomenon. We have focused too much on climate change as a physical phenomenon, largely ignoring the ways in which the idea of climate change gets constructed and appropriated for use. Such a purely physical framing invites a problem-solution dialectic. But if climate change is an idea as much as it is a physical reality, is it possible to solve it any more than one can 'solve' violence?

The Hot Topic defines the problem of climate change through the lens of natural science and offers solutions that borrow largely from science, engineering and technology. Framing it differently as an idea would instead force us to explore our values, our relationships and our view of ourselves. Is our purpose on Earth to secure greater affluence, to seek justice, or is it merely to survive? This is the really big question that the idea of climate change is demanding an answer to; and on this *The Hot Topic* has too little to say.

39 What was the Copenhagen climate change conference really about?

13 March 2009

This essay was written for the American *SeedMagazine*[113]

> It's problematic when largely unresolved debates among the world's climate change researchers get reduced to six key messages

The largest academic conference that has yet been devoted to the subject of climate change finished yesterday in Copenhagen. Between 2,000 and 2,500 researchers from around the world attended three days of meetings during which 600 oral presentations (together with several hundred posters on display) were delivered on topics ranging from the ethics of energy sufficiency to the role of icons in communicating climate change to the dynamics of continental ice sheets.

I attended the Conference, chaired a session, listened to several presentations, read a number of posters, and talked with dozens of colleagues from around the world. The breadth of research on climate change being presented was impressive, as was the vigour and thoughtfulness of the informal discussions being conducted during coffee breaks, evening receptions, and side-meetings.

What intrigued me most, however, was the final conference statement issued yesterday, a statement drafted by the conference's Scientific Writing Team. It contained six key messages and was handed to the Danish Prime Minister Mr Anders Fogh Rasmusson. The messages focused, respectively, on Climatic Trends, Social Disruption, Long-term Strategy, Equity Dimensions, Inaction is Inexcusable, and Meeting the Challenge. A fuller version of this statement will

---

113 Hulme, M. (2009) What was the Copenhagen climate change conference really about? *SeedMagazine* 13 March, seedmagazine.com/content/article/what_was_the_copenhagen_climate_change_conference_really_about/. A version of this essay also appeared on the *BBC News On-line* Green Room, 16 March 2009. Together with some social science colleagues, some of these thoughts were published as correspondence in the journal *Science*: Hulme, M., Boykoff, M., Gupta, J., Heyd, T., Jaeger, J., Jamieson, D., Lemos, M.C., O'Brien, K., Roberts, T., Rockstrom, J. and Vogel, C. (2009) Conference covered climate from all angles *Science* 324, 881–882.

be prepared and circulated to key negotiators and politicians ahead of the 15th Conference of the Parties (COP15) to the UN Framework Convention on Climate Change to be held in December this year, also in Copenhagen.

The conference, and the final conference statement,[114] has been widely reported as one at which the world's scientists delivered a final warning to climate change negotiators about the necessity for a powerful political deal on climate change to be reached at COP15 (some commentators have branded it the 'Emergency Science Conference'). The key messages include statements that 'the worst-case IPCC scenario trajectories (or even worse) are being realised,' that 'there is no excuse for inaction,' that 'the influence of vested interests that increase emissions' must be reduced, and that 'regardless of how dangerous climate change is defined,' rapid, sustained, and effective mitigation is required to avoid reaching it.

There is a fair amount of 'motherhood and apple pie' involved in the 600-word statement – who could disagree, for example, that climate risks are felt unevenly across the world or that we need sustainable jobs. But there are two aspects of this statement which are noteworthy and on which I would like to reflect: Whose views does the statement represent and what are the actions being called for?

The Copenhagen Climate Change Congress was no IPCC. This was not a process initiated and conducted by the world's governments, there was no systematic synthesis, assessment, and review of research findings as in the IPCC, and there was certainly no collective process for the 2,500 researchers gathered in Copenhagen to consider drafts of the six key messages nor to offer their own suggestions for what politicians may need to hear. The conference was in fact convened by no established academic or professional body. Unlike the American Geophysical Union, the World Meteorological Organisation or the UK's Royal Society – who also hold large conferences and who from time-to-time issue carefully-worded statements representing the views of professional bodies – this conference was organised by the International Alliance of Research Universities (IARU), a little-heard-of coalition launched in January 2006 consisting of ten of the world's self-proclaimed elite universities, including of course the University of Copenhagen.

IARU is not accountable to anyone and has no professional membership. It is not accountable to governments, to professional scientific associations, nor to international scientific bodies operating under the umbrella of the UN. The conference statement therefore simply carries the weight of the Secretariat of this ad hoc conference, directed and steered by ten self-elected universities. The six key messages are not the collective voice of 2,500 researchers, nor are they the voice of established bodies such as the World Meteorological Organisation. Neither are they the messages arising from a collective endeavour of experts, for example through a considered process of screening, synthesising, and reviewing of the knowledge presented in Copenhagen this week. They are instead a set of

114 Available here: http://climatecongress.ku.dk/newsroom/congress_key_messages/ (accessed 22 December 2012).

messages drafted largely before the conference started by the organising committee, sifting through research that they see emerging around the world and interpreting it for a political audience.

Which leads me to the second curiosity about this conference statement, what exactly is the 'action' the conference statement is calling for? Are these messages expressing the findings of science or are they expressing political opinions? I have no problem with scientists offering clear political messages as long as they are clearly recognised as such. And the conference chair herself, Professor Katherine Richardson, has described the messages as politically motivated. All well and good.

But then we need to be clear about what authority these political messages carry. They carry the authority of the people who drafted them – and no more. Not the authority of the 2,500 expert researchers gathered at the conference. And certainly not the authority of collective global science. Caught between summarising scientific knowledge and offering political interpretations of such knowledge, the six key messages seem rather ambivalent in what they are saying. It is as if they are not sure how to combine the quite precise statements of science with a set of more contested political interpretations.

Which brings us back to the calls for action and the 'inexcusability of inaction.' What action on climate change exactly is being called for? During the conference there were debates amongst the experts about whether a carbon tax or carbon trading is the way to go. There were debates amongst the experts about whether or not we should abandon the 'two degrees' target as unachievable. There were debates about whether or not a portfolio of geo-engineering strategies now really needs to start being researched and promoted. And there were debates about the epistemological limits to model-based predictions of the future. There were debates about the role of behavioural change versus technological change, about the role of religions in mitigation and adaptation, and about the forms of governance most likely to deliver carbon reductions.

These are all valid debates to have. And they were debates that did occur during the conference. Experts from the natural sciences and social sciences, from engineering and policy sciences, from economics and the humanities, all presented findings from their work and these were discussed and argued over. These debates mixed science, values, ethics, and politics. This is the reality of how climate change now engages with the worlds of theoretical, empirical, and philosophical investigation.

It therefore seems problematic to me when such lively, well-informed, and yet largely unresolved debates among a substantial cohort of the world's climate change researchers get reduced to six key messages, messages that on the one hand carry the aura of urgency, precision, and scientific authority – 'there is no excuse for inaction' – and yet at the same time remain so imprecise as to resolve nothing in political terms.

In fact, we are no further forward after the Copenhagen Congress this week than before it. All options for attending to climate change – all political options – are, rightly, still on the table. Is it to be a carbon tax or carbon trading? Do we stick with 'two degrees' or abandon it? Do we promote geo-engineering or do we not? Do we coerce lifestyle change or not? Do we invest in direct poverty alleviation or in the Green New Deal?

A gathering of scientists and researchers has resolved nothing of the politics of climate change. But then why should it? All that can be told – and certainly should be told – is that climate change brings new and changed risks, that these risks can have a range of significant implications under different conditions, that there is an array of political considerations to be taken into account when judging what needs to be done, and there are a portfolio of powerful, but somewhat untested, policy measures that could be tried.

The rest is all politics. And we should let politics decide without being ambushed by a chimera of political prescription dressed up as (false) scientific unanimity.

40 Climate change: no Eden, no apocalypse

5 September 2009

An essay written for *New Scientist* magazine[115]

Climate change is everywhere. Not only is the physical climate changing, but the idea of climate change is now active across the full range of human endeavours. Climate change has moved from being a predominantly physical phenomenon to being a social one, in the process reshaping the way we think about ourselves, our societies and humanity's place on Earth.

I am primarily a climate scientist who has worked with climate data, models and scenarios. But I am now more interested in how we think and talk about climate change, how we use the idea to support various projects, and how – paradoxically – we could use it to make the world a better place. I argue that just as we need to understand the physical changes that are sweeping the planet, we also need to understand climate change as a cultural and psychological phenomenon.

And it is a phenomenon. Just as the transformation of the physical climate is inescapable, so the idea of climate change is now unavoidable. It is circulating anxiously in the worlds of domestic politics and international diplomacy, and with mobilising force in business, law, academia, development, welfare, religion, ethics, art and celebrity.

Yet in each of these spheres the idea of climate change carries quite different meanings and seems to imply different courses of action. The Intergovernmental Panel on Climate Change has constructed a powerful scientific consensus about the physical transformation of the world's climate. This is a reality that I believe in. But there is no comparable consensus about what the idea of climate change actually means. If we are to use the idea constructively, we first need new ways of looking at the phenomenon and making sense of it.

115 Hulme, M. (2009) Climate change: no Eden, no apocalypse *New Scientist* 5 September 28–29.

One way I do this is to rethink our discourses about climate change in terms of four enduring myths. I use 'myths' not to imply falsehoods but in the anthropological sense – stories we tell that embody deeper assumptions about the world around us. First is the Edenic myth, which talks about climate change using the language of lament and nostalgia, revealing our desire to return to some simpler, more innocent era. In this myth, climate is cast as part of a fragile natural world that needs to be protected. It shows that we are uneasy with the unsought powers we now have to change the global climate.

Next, the Apocalyptic myth talks about climate in the language of fear and disaster. This myth reveals our endemic worry about the future, but also acts as a call to action. Then there is the Promethean myth, named after the Greek deity who stole fire from Zeus and gave it to the mortals. This talks about climate as something we must control, revealing our desire for dominance and mastery over nature but also that we lack the wisdom and humility to exercise it. Finally, the Themisian myth, named after the Greek goddess of natural law and order, talks about climate change using the language of justice and equity. Climate change becomes an idea around which calls for environmental justice are announced, revealing the human urge to right wrongs.

The value in identifying these mythical stories in our discourses about climate change is that they allow us to see climate change not as simply an environmental problem to be solved, but as an idea that is being mobilised in various ways around the world. If we continue to naively understand the climate system as something to be mastered and controlled, then we will have missed the main opportunities offered us by climate change.

From a practical perspective, that means rethinking our responses to climate change. Rather than placing ourselves in a 'fight against climate change' we should use the idea of climate change to rethink and renegotiate our wider social and political goals. How so? For one thing, climate change allows us to examine our projects more closely and more honestly than we have been used to, whether they be projects of trade, community-building, poverty reduction, demographic management, social and psychological health, personal well-being or self-determination. Climate change demands that we focus on the long-term implications of our short-term choices and recognise the global reach of our actions. This means asking both 'what is the impact of this project on the climate?' and also 'how does the reality of climate change alter how we can achieve this goal?'

Climate change also teaches us to rethink what we really want for ourselves and humanity. The four mythical ways of thinking about climate change reflect back to us truths about the human condition that are both comforting and disturbing. They suggest that even were we to know precisely what we wanted – wealth, communal harmony, social justice or mere survival – we are limited in our abilities to acquire or deliver those goals.

Having established that climate change is as much an idea as a physical phenomenon, we can deploy it in positive and creative ways. It can stimulate new thinking about technology. It can inspire new artistic creations. It can provoke new ethical and theological thinking. It can arouse new interest in how science and culture interrelate. It can galvanise new social movements to explore new ways of living in urban and rural settings. It can touch each one of us as we reflect on the goals and values that matter to us. And, of course, the idea of climate change can invigorate efforts to protect ourselves from the hazards of climate change.

It is important to note that these creative uses of the idea of climate change do not demand consensus over its meaning. Indeed, they may be hindered by the search for agreement. They thrive in conditions of pluralism. Nor are they uses that will necessarily lead to stabilising climate – they will not 'solve' climate change. This does not imply passivity in the face of change, however. Nor does it allow us to deny that our actions on this planet are changing the climate. But it does suggest that making climate control our number one political priority might not be the most fruitful way of using the idea of climate change.

The world's climates will keep on changing, with human influences now inextricably entangled with those of nature. So too will the idea of climate change keep changing as we find new ways of using it to meet our needs. We will continue to create and tell new stories about climate change and mobilise these stories in support of our projects. Whereas a modernist reading of climate may once have regarded it as merely a physical condition for human action, we must now come to terms with climate change operating simultaneously as an overlying, but more fluid, imaginative condition of human existence.

41 Heated debate

Spring 2010

Essay written for the *Journal* of the
Royal Society of Arts[116]

> There's no such thing as right and wrong when it comes to tackling
> climate change, says Mike Hulme. That's why we need to stop
> looking for scapegoats and start engaging in honest discussion.

One of the enduring characteristics of public debates and political negotiations
about climate change is that the protagonists end up arguing about different
things. Political arguments masquerade as arguments about science; ethical
arguments become economic ones. Legitimate differences about ideologies and
values are reduced to trading blows about the 'right' numbers – the decimal points
on rates of warming; the number of noughts in the cost of climate change. We are
not being honest with one another. The consequence is that the quality of both
science and public debate suffers.

Since it first emerged as a prominent public policy issue in the late 1980s,
anthropogenic climate change has evolved into an idea that now carries an
astonishing amount of ideological freight. Yet, too often, arguments about climate
change continue to treat it as an environmental problem to be solved. But climate
change is not a phenomenon of this kind. It is not like mercury pollution in rivers,
asbestos in buildings or even ozone-depleting gases entering the stratosphere.
These relatively 'tame' problems lend themselves to relatively straightforward
solutions: the Montreal Protocol, for example, which opened for signature in
1987, successfully restricted and then prohibited the use of ozone-depleting
substances.

Not so with climate change. Climate change is a 'wicked' problem. There is no
unambiguous formulation of what the problem is and no opportunity to learn
from other, similar cases. Proposed solutions are so embedded in matrices of
social, economic and political cause and effect that they are likely to spawn

116 Hulme, M. (2010) Heated debate *RSA Journal* Spring 36–37 http://www.thersa.org/
fellowship/journal/archive/winter-2010/features/heated-debate.

further unforeseeable and unwelcome side effects. It is not surprising that some have despaired and are now suggesting that we must cut through this Gordian knot and find a more direct solution through climate engineering. Pumping aerosols into the stratosphere, it is claimed, would allow us to control the planet's temperature directly, bypassing these troublesome entanglements and this social inertia.

But climate change has come to signify far more than the physical ramifications of human disturbance to the composition of the Earth's atmosphere and its energy balance. Climate change has become as much a social phenomenon as it is a physical one. Arguments about the causes and consequences of climate change – and the solutions to it – have become nothing less than arguments about some of the most intractable social, ethical and political disputes of our era: the endurance of chronic poverty in a world of riches; the nature of the social contract between state and citizen; the cultural authority of scientific knowledge; and the role of technology in delivering social goods. Climate change has become a metaphor for the imagined future of human life and civilisation on Earth.

The different meanings that can be attached to the idea of climate change are illustrated well by considering ways in which the issue is framed in India. For many in this country, the key concerns are how to secure financial reparations for environmental damage caused by northern nations through the proxy of climate and how to use climate change to advance the development of the 500 million people living in absolute poverty. This framing of climate change is very different from that which prevails in much western discourse and implies a very different set of international and domestic policy prescriptions. The issue is less about how to reverse a two-degree temperature change, how to save polar bears or how to avoid metaphorical tipping points than it is about how to secure hundreds of billions of dollars to invest in basic human welfare.

It is not surprising, then, that arguments about climate change are invested with powerful ideological instincts and interests. Solutions to climate change vary from market-based mechanisms and technology-driven innovation to justice-focused initiatives and low-consumption localism as a form of lifestyle, each carrying ideological commitments. It is despairingly naive to reduce such intense (and legitimate) arguments to the polarities of 'belief' or 'scepticism' about science.

Belief in what, exactly? Is it the belief that humans are contributing significantly to climate change? Yes, science can speak authoritatively on this question. Or a belief that the possible consequences of future change warrant an emergency policy programme? Scientific evidence here offers only one strand of the necessary reasoning. Or a belief that such an emergency policy programme must be secured through an international, legally binding targets-and-timetables approach, such as Kyoto? On this, science has very little to say.

On the other hand, what exactly is it that the so-called sceptics are charged with? Scepticism that environmental scientists, businesses and central government are in collusion to fabricate evidence? This is barely plausible. Or scepticism that claims about the future that are based on scientific knowledge are sometimes overstretched and underplay uncertainties? The latter is a warning that all would do well to heed. The problem here is the tendency to reduce all these complexities into a simple litmus test of whether or not someone believes orthodox scientific claims about the causes and consequences of climate change. This is dividing the world into goodies and baddies, believers and deniers. Climate change demands of us something much more sophisticated than this.

Rather than reducing climate change to arguments about how settled – or not – the science is (predictions in environmental science are rarely, if ever, settled), we need to provide the intellectual, educational, ethical and political spaces to argue fearlessly with one another about the very things that the idea of climate change demands we take a position on. These include our attitudes to global poverty, the role of the state in behavioural change, the tension between acting on knowledge or on uncertainty, the meaning of human security and the value of technological innovation. Where we stand on issues such as these will determine which sort of solutions to climate change we choose to advocate.

None of these things is new. They have been around for a long time – for at least the 50 years since novelist and physicist C.P. Snow declared that advancing science and technology was the only sure way to secure human welfare.

But the idea of climate change – suggesting, as it does, that our current development trajectory may not be as systemically benign as we might wish – demands that we re-examine these troubling issues. We must examine them explicitly and honestly. And we must respect the different legitimate positions people adopt about these ideological and ethical entanglements when they appear in public spaces. Indeed, we must foster such exchanges without applying the sleight of hand that turns them back into arguments about belief (or otherwise) in scientific claims.

Neither scientists nor politicians should try to discredit unorthodox views about how to respond to climate change by using the pejorative labels of 'denialist' or 'flat-earther'. Scientists must learn to respect their public audiences and to listen more closely to them. Now is no time for the elite to despair of democracy. We have only one planet, but we also have only one political system that most people would choose to live under. Politicians must learn not to hide behind science when asked to make complex judgements. Science is useful as a form of systematic critical enquiry into the functioning of the physical world, but it is not a substitute for political judgement, negotiation and compromise.

42 You've been framed

5 July 2011

An essay written for the Australian
on-line magazine *The Conversation*[117]

Six new ways to understand climate change

There are many ways to frame the phenomenon of climate change. Some may be more engaging and some more helpful than others. Some may play looser with the facts. And yet no frames – even those that remain faithful to the facts – can be entirely neutral with respect to the effects that they generate on their audiences. Take the opening item in *The Conversation*'s recent climate change series 'Clearing up the Climate Debate'.[118] This open letter boldly states its framing narrative: 'The overwhelming scientific evidence tells us that human greenhouse gas emissions are resulting in climate changes that cannot be explained by natural causes. Climate change is real, we are causing it, and it is happening right now.'

Fact. Nothing to challenge there. But how about this alternative?

'The overwhelming scientific evidence tells us that human greenhouse gas emissions, land use changes and aerosol pollution are all contributing to regional and global climate changes, which exacerbate the changes and variability in climates brought about by natural causes. Because humans are contributing to climate change, it is happening now and in the future for a much more complex set of reasons than in previous human history.'

I'm confident too that none of my climate science colleagues would find anything to challenge in this statement. And yet these two different provocations – two different framings of climate change – open up the possibility of very different forms of public and policy engagement with the issue. They shape the response. The latter framing, for example, emphasises that human influences on climate are not just about greenhouse gas emissions (and hence that climate

117 Hulme, M. (2011) You've been framed: six new ways to understand climate change *The Conversation* http://theconversation.edu.au/youve-been-framed-six-new-ways-to-understand-climate-change-2119.

118 The essays to which I was partly responding can be found here: http://theconversation.edu.au/clearing-up-the-climate-debate-2078 (accessed 22 December 2012).

change is not just about fossil energy use), but also result from land use changes (emissions and albedo effects) and from aerosols (dust, sulphates and soot). It emphasises that these human effects on climate are as much regional as they are global. And it emphasises that the interplay between human and natural effects on climate are complex and that this complexity is novel.

The frame offered by the 87 Australian academics who signed the 'open letter' is more partial than mine and also, I suggest, is one which is (perhaps deliberately) more provocative. It may work well if their intention is to reinforce the polarisation of opinion that exists around climate change science or if they are using scientific claims to justify a particular set of policy interventions.

Yet there are important aspects of scientific knowledge about the climate system that are accommodating to more nuanced interpretations of uncertainty and which open up more diverse sets of policy strategies. It is these aspects which my framing is seeking to foreground. My general point then is that how one frames a complex issue – and we all agree that climate change is complex – inevitably emphasises some aspects of that issue while de-emphasising others. And that these emphasis effects are not neutral. They result from judgements – whether careful or careless – made by those framing the issue and they have significant consequences for how audiences receive and engage with the communication.

Framing effects around climate change are very powerful. My recent speaking tour of Australia and my book *Why We Disagree About Climate Change* focused on this – and why it matters. In particular, I suggested six powerful frames through which climate change is presented in public discourse:

- climate change as market failure
- as technological risk
- as global injustice
- as over consumption
- as mostly natural
- as planetary 'tipping point'.

Framing climate change as market failure draws attention to a particular set of policy interventions: those which seek to 'correct' the market by introducing pricing mechanisms for greenhouse gases.

Climate change when framed as a 'manufactured risk' focuses on the inadvertent downsides of our ubiquitous fossil-energy based technologies. It lends itself to a policy agenda which promotes technology innovation as the solution to climate change.

Radically different, however, is the frame of global injustice. Here, climate change is presented as the result of historical and structural inequalities in access to wealth and power and hence unequal life chances. Climate change is all about the rich and privileged exploiting the poor and disadvantaged. Any solutions to climate change that fail to tackle that underlying 'fact' are doomed to fail.

A related frame, but one with a different emphasis, is climate change as the result of overconsumption: too many (rich) people consuming too many (material) things. If this is the case then policy interventions need to be much more radical than simply putting a price on carbon or promoting new clean energy technologies. The focus should be on dematerialising economies or else on promoting fertility management.

A fifth frame would offer climate change as being mostly natural. Human influences on the global climate system can only be small relative to nature and so the emphasis should be less on carbon and energy policy and more about adaptation: enabling societies to cope with climate hazards irrespective of cause.

Last is the frame of planetary 'tipping point' which has arisen since 2005. Climate change carries with it the attendant dangers of pushing the planetary system into radically different states. Such 'tipping points' may be reached well before carbon markets, clean energy or economic de-growth will be attained and so new large-scale climate intervention technologies – a so-called Plan B – need to be developed and put on stand-by.

These six frames around climate change all attract powerful audiences, interests and actors in their support. All of them – with the exception of climate change as mostly natural – would be broadly consistent with the scientific knowledge assessed by the Intergovernmental Panel on Climate Change. And yet because they are rooted in different ideologies and views of the relationship between humans, technologies and nature they filter and interpret that scientific evidence in different ways and use it to justify certain forms of policy.

The human influences on climate change – and the policy significance of these influences – are too complex to reduce public debate around climate change to a bi-polar caricature: mainstream scientists versus sceptics; believers versus deniers; liberal progressives versus conservatives. I have shown that there are multiple framings of climate change – and there are more than I've offered here – in which scientific evidence, attitudes to risk, political ideology, myths of nature and so on are deeply interweaved and entangled. It is deeply inconvenient I know, but there is no single rational response to the fact that we are an agent powerfully shaping the planet.

We all need to be aware of our own framings, our own preferences, beliefs and ideologies, when it comes to debating what we should do.

Section seven

Controversy

Introduction

In the winter of 2009/10 two significant controversies took centre-stage in public forums dealing with climate change, at least they did so in Anglophone countries. The first of these was quickly dubbed 'Climategate', and revolved around the interpretation and significance of professional email correspondence between leading climate scientists. The second, a few weeks later, concerned the accuracy and authority of the UN's Intergovernmental Panel on Climate Change (IPCC). Sandwiched in between these two events was the 15th Conference of the Parties to the UN Framework Convention on Climate Change (COP15) which took place at Copenhagen.

The articles selected for this section on Controversy have as their reference point these two remarkable episodes. But the first item included here (Chapter 43) pre-dates Climategate and is the account of an interview Stuart Blackman conducted with me for the on-line news magazine *The Register*. It was prompted by the publication in April 2009 of my book *Why We Disagree About Climate Change* and the interview offers an insight into why I might well have expected the sorts of controversies seen a few months later.

The emails acquired from the back-up server of the Climatic Research Unit (CRU) in the School of Environmental Sciences at the University of East Anglia (UEA) were posted on-line on Thursday 19 November 2009 and within 24 hours the story had gone viral on the internet. The UK and US mainstream media quickly followed and the battle over who were the heroes and villains of Climategate raged intensely for several weeks. I found myself in an interesting position. I was a peripheral actor in the controversy – some of my emails dating back to the late 1990s were included in the released batch and I was labelled by some commentators as a member of 'The Team', a pejorative term alluding to those scientists associated with the so-called 'hockey-stick curve' (in fact I have never worked or published on paleoclimate data in my career). I was also party to some information which had a bearing on some of the criticisms being levelled at my ex-colleagues in CRU. At the same time I was an employee of UEA, which found itself having to answer difficult questions and which quickly moved to issue for its staff rules of engagement with the media.

I had a strong view about the role of climate science in public and policy debates on climate change, a view which had evolved over the previous decade (see Chapters 30 and 31) and which was at odds with some of my ex-colleagues. As my interview with *The Register* shows, I did not feel that the framing of climate change, nor the role of climate science in this framing, was serving my own sense of a healthy democracy. Following a previously-arranged meeting with Jerry Ravetz in Oxford on 24 November we decided that we would try and publish a commentary on the wider significance of Climategate. This commentary eventually found a home with the *BBC News On-line* Green Room and is included here as Chapter 44. Around the same time I was approached by the London office of *The Wall Street Journal* to write an op-ed for that newspaper and this is included here as Chapter 45.

Both of these items – which appeared on consecutive days in December 2009 – required authorisation by my head of school and by the university's media office, an indication of the unusual circumstances in which the University of East Anglia then found itself. The BBC piece was a commentary on the nature of science and scientific evidence on matters of high public policy significance, drawing upon Jerry's idea of post-normal science (see Chapter 17). *The Wall Street Journal* op-ed was a chance to further expose my argument of *Why We Disagree About Climate Change* – the need to move the locus of disagreement about climate change away from science and onto the territory of beliefs, ideologies and values.

Following the debacle of the Copenhagen negotiations, a few weeks into the new year of 2010 saw the spotlight move onto the IPCC. This was triggered by an erroneous claim in its Fourth Assessment Report about the melting of Himalayan glaciers. I have included two articles I wrote about this controversy, one for the widely read on-line portal *SciDev.net* (Chapter 46) and one for the comment pages of the on-line journal *Environmental Research Letters* (Chapter 48). The former focuses on the poor handling of the controversy by the IPCC management, and particularly by its chairman, which compounded the original problem. The latter essay comments more widely on the nature of the IPCC, how it gained and lost authority and on the need for its reformation.

Sandwiched in between these two essays on the IPCC is the unpublished transcript of an interview (Chapter 47) which I offered at the end of June 2010 to a MSc student in Science Communication at Imperial College, London. This was part of her dissertation research which was to be a 30-minute video documentary on climate science and scientific controversy. I have included this interview in this section since it ranges widely over the relationships between climate science, public perceptions and climate politics, framed by the controversies of Climategate and the IPCC.

The last two items in this section return to the Climategate controversy. The first (Chapter 49) is an essay published in *The Guardian* newspaper in November 2010, exactly 12 months after the CRU emails were first made public. I draw out three changes which I had observed taking place during this period: changes in the practices of climate science, in the public framing of climate change and in the new emerging pragmatism around climate policy. Controversies are always moments for learning and change and this was undoubtedly true in the case of

Climategate which is a theme I take up in Chapter 50, an unpublished essay written specially for this book. Here I consider the questions: Have the consequences of Climategate been good or bad; and for whom? How has climate science changed as a result? And how has the imaginative force of climate change as an idea changed since the autumn of 2009?

43 Top British boffin: time to ditch the climate consensus

6 May 2009

An interview with Stuart Blackman for
The Register[119]

Don't use science to get round politics, says Hulme

Just two years ago, Mike Hulme would have been about the last person you'd expect to hear criticising conventional climate change wisdom. Back then, he was the founding director of the Tyndall Centre for Climate Change Research, an organisation so revered by environmentalists that it could be mistaken for the academic wing of the green movement. Since leaving Tyndall – and as we found out in a telephone interview – he has come out of the climate change closet as an outspoken critic of such sacred cows as the UN's IPCC, the 'consensus', the over-emphasis on scientific evidence in political debates about climate change, and to defend the rights of so-called 'deniers' to contribute to those debates.

As Professor of Climate Change at the University of East Anglia, Hulme remains one of the UK's most distinguished and high-profile climate scientists. In his new book, *Why We Disagree About Climate Change*, he explores how the issue of climate change has come to be such a dominant issue in modern politics. He treats climate change not as a problem that we need to solve – indeed, he believes that the complexity of the issue means that it cannot be solved, only lived with – and instead considers it as much of a cultural idea as a physical phenomenon.

Perhaps the most surprising thing to hear from a climate scientist writing about climate change is that climate science has for too long had the monopoly in climate change debates. When we spoke to him on the phone, Hulme cited as evidence the 2007 protests against Heathrow's third runway, where marchers made their case by waving a research paper at the TV cameras under a banner bearing the slogan 'We are armed only with peer reviewed science'. (The paper wasn't actually peer-reviewed).

---

119 Interview with Stuart Blackman of *The Register*, 6 May 2009: 'Top British boffin: time to ditch the climate consensus' http://www.theregister.co.uk/2009/05/06/mike_hulme_interview/.

'To me, that's the most dispiriting position,' says Hulme. 'For these people who feel so passionately about this, their ultimate authority is a report from a group of scientists, and they're saying "this is where we stand, forget about our moral concerns, forget about our ethical positions, forget about whether we are Right, Left or centre, forget about whether we are Christians or Buddhists, no, none of that matters." The only thing that matters is that they're holding a report from peer-reviewed science that in itself justifies their position.'

'Uncertainty, and things like that'

And it's not just protesters who are hiding behind the authority of science. World leaders are doing it, too. Hulme despairs over the comments made to the Copenhagen Climate Change Congress in March by Anders Fogh Rasmussen, then the Danish Prime Minister. Rasmussen told delegates that 'science should be the basis for decision-making in this field', and asked scientists to keep it simple, 'not to provide us with too many moving targets ... and not too many considerations on uncertainty and risk and things like that.'

'That's just classic,' says Hulme. 'Here's this politician telling the scientists "we can't do this without you. Give us the numbers. But by the way, make them simple, and make them precise".' Hulme believes that this dependence of politics on science expects too much of science's ability to explain and to predict, and that this is a burden that science cannot carry. Science is exposing its vulnerabilities, he says. And in overselling itself, the risks are very substantial. 'It's like the classic case of the dodgy dossier'.

Making politics disappear

He stresses that he has little problem with the basic scientific understanding of climate change. It's just that, if progress is to be made in debates on how to respond to that knowledge, they need to be opened up to other disciplines, from the arts and humanities, for example – and to good old-fashioned politics and ideologies.

'However much we agree on the fundamentals of the physics of climate change, there are huge ethical, political and ideological differences that remain about what climate change signifies for society', says Hulme. 'And if one pretends that we can gloss over those, converging on a single political position, where there is no party political debate and differentiation, then we're losing some of the essential dimensions of climate change that we have to engage with. It narrows down debate rather than opening it up.'

Which is why Hulme has opposed the idea of UK cross-party consensus. 'Climate change can only be understood from a position of dissensus, rather than artificially solved by creating consensus,' he says. Similarly, while he is sympathetic to the ambitions of the Inter-Governmental Panel on Climate Change (IPCC), he is critical of the way it is widely cited as the last word on climate issues. 'It's won the Nobel Peace Prize for goodness' sake – it can't be challenged!'

He also regards the IPCC as too selective in terms of both the geographical regions from which it draws its knowledge and in its academic scope. 'It is hugely dominated by the natural sciences, economics and engineering. The social sciences hardly get a look in, and the humanities none at all. For example, it does not include anthropological understandings of weather and climate or any historical perspectives on how societies and climates have interacted historically.'

'If climate change is the biggest issue facing the future of human civilisation, to use the rhetoric, then surely a body charged to assess what humans know about climate change should actually be assessing all forms of knowledge.' Moreover, says Hulme, no one is even quite sure what sort of knowledge it is that the IPCC, as a 'boundary organisation' – part science, part politics – actually produces. Nor how the world at large interprets that hybrid knowledge. Even more fundamentally, he says, it is far from clear that the IPCC has actually allowed us to do 'better science': 'Or has it actually narrowed the way we frame and ask questions in climate change research?' Hulme wonders.

No denier

And yet even though the IPCC is an institutional experiment as much as a scientific one – and despite its occupying a position of huge influence in the world – few sociologists seem to be scrutinising its workings. This may in part be due to the fact that action on climate change is widely seen as a progressive goal, says Hulme, and being a generally progressive sort of bunch, social scientists might be reticent to impede proceedings, or to be seen to give succour to right-wing 'denialists'.

'That's an accusation that has been charged at me, that I'm simply lending ammunition to people who generally are politically conservative, and who want to discredit the basic physics behind climate change.' In pushing to open up climate change debates to non-scientific disciplines, Hulme runs the risk perhaps of attracting accusations of not only 'denier', but also of 'relativist', which is almost as dirty a word in scientific circles. Hulme's Christian beliefs might be a further invitation to ad hominem responses.

But any attacks that were aimed at him on these grounds would demonstrate his point nicely. After all, much of the abuse that is hurled across the climate divide comes from those who like to believe that it is *they* who are dealing in a currency of proper science – bias and ideology is what the opposition does. Hence the vitriol aimed at Bjørn Lomborg over the years.

'It was interesting as to why he received such hate-mail from very well respected academics rather than simply engaging in the arguments,' says Hulme. 'It became very very heavily and easily personalised, when actually Lomborg's position is an entirely defendable position. I mean, you can disagree with it, and you can find flaws in his argument, but let's find those flaws and let's have a disagreement, rather than suddenly becoming reactionaries overnight. And I think there's too much of that. And it's an interesting question as to why it is that people feel that climate change is somehow the issue beyond all other issues today that one has to stand on shoulder to shoulder and not allow any chink in because it would allow the powers of darkness to somehow gain the upper hand.'

For Hulme, for open debate to be possible, there must be a recognition on all sides that we all bring a host of values, beliefs and influences to the table along with our knowledge, expertise and training.

'If, say, Jim Hansen or Fred Singer and I sat down and looked at the same scientific evidence, we would come up with a very different set of prescriptions. Now, why is that? Is it because our scientific training is deficient, and he's seeing more than I'm seeing, or I'm seeing more than he's seeing? I don't think it is. I think actually there's a lot of stuff that's going on here. And that's actually what we have to get down to – to root out, and expose, accept, and work within these broader, deeper sources of disagreement.'

'To hide behind the dubious precision of scientific numbers, and not actually expose one's own ideologies or beliefs or values and judgements is undermining both politics and science', says Hulme.

A bigger debate

That his thesis has the potential to draw in and engage disparate climate change factions is suggested by the cover-blurb testimonies from an oil company advisor, a deep ecologist, a sociologist, and an environmental scientist. But Hulme has his work cut out. Even as he spoke to us, President Obama was declaring in his address to the US National Academy of Sciences: 'Under my administration, the days of science taking a back seat to ideology are over'.

It's not hard to get labelled a climate change 'denier'. You don't even have to deny that climate change is real, man-made and a problem. As Bjørn Lomborg, climatologist Patrick Michaels and political scientist Professor Roger Pielke Jr. have discovered, you merely have to challenge the orthodox political policy responses. Or, like Climate Audit's Steve McIntyre, dare to scrutinise the statistical workings behind influential climate research papers. If you stray from agreeing with the political prescription, you're an immoral person.

So, how long, we wonder, before Mike Hulme attracts the same accusation?

44 'Show your working': what Climategate means

1 December 2009

An essay written with Jerry Ravetz for the *BBC News On-line* Green Room[120]

> The 'Climategate' affair is being intensely debated on the web. But what does it imply for climate science? Here, Mike Hulme and Jerome Ravetz say it shows that we need a more concerted effort to explain and engage the public in understanding the processes and practices of science and scientists.

As the repercussions of Climategate reverberate around the virtual community of global citizens, we believe it is both important and urgent to reflect on what this moment is telling us about the practice of science in the twenty-first century. In particular, what is it telling us about the social status and perceived authority of scientific claims about climate change?

We argue that the evolving practice of science in the contemporary world must be different from the classic view of disinterested – almost robotic – humans establishing objective claims to universal truth. Climate change policies are claimed to be grounded in scientific knowledge about physical cause and effect and about reliable projections of the future. As opposed to other ways of knowing the world around us – through intuition, inherited belief, myth – such scientific knowledge retains its authority by widespread trust in science's reassuring norms of objectivity, universality and disinterestedness. These perceived norms work to guarantee to the public trustworthy scientific knowledge, and allow such knowledge to claim high authority in political deliberation and argumentation; this, at least, is what historically has been argued in the case of climate change.

What distinguishes science from other forms of knowledge? On what basis does scientific knowledge earn its high status and authority? What are the minimum standards of scientific practice that ensure it is trustworthy? For an open, enquiring and participative society, these are questions that have become

120 Hulme, M. and Ravetz, J. (2009) 'Show your working': what Climategate means *BBC News On-line* Green Room 1 December http://news.bbc.co.uk/1/hi/sci/tech/8388485.stm.

much more important in the wake of Climategate. They are also questions that scientists should continually be asking of themselves as the political and cultural worlds within which they do their work rapidly change. Doing science in 2010 demands something rather different from scientists than did science in 1960, or even in 1985.

How science has evolved

The understanding of science as a social activity has changed quite radically in the last 50 years. The classic virtues of scientific objectivity, universality and disinterestedness can no longer be claimed to be automatically effective as the essential properties of scientific knowledge. Instead, warranted knowledge – knowledge that is authoritative, reliable and guaranteed on the basis of how it has been acquired – has become more sought after than the ideal of some ultimately true and objective knowledge. Warranted knowledge places great weight on ensuring that the authenticating roles of socially agreed norms and practices in science are adequately fulfilled – what in other fields is called quality assurance. And science earns its status in society from strict adherence to such norms.

For climate change, this may mean the adequate operation of professional peer review, the sharing of empirical data, the open acknowledgement of errors, and openness about one's funders. Crucially, the idea of warranted knowledge also recognises that these internal norms and practices will change over time in response to external changes in political culture, science funding and communication technologies. In certain areas of research – and climate change is certainly one of these – the authenticating of scientific knowledge now demands two further things: an engagement with expertise outside the laboratory, and responsiveness to the natural scepticism and desire for scrutiny of an educated public. The public may not be able to follow radiation physics, but they can follow an argument; they may not be able to describe fluid dynamics using mathematics, but they can recognise evasiveness when they see it.

Where claims of scientific knowledge provide the basis of significant public policy, demands for what has been called 'extended peer review' and 'the democratisation of science' become overwhelming. Extended peer review is an idea that can take many forms. It may mean the involvement of a wider range of professionals than just scientists. The Intergovernmental Panel on Climate Change (IPCC), for example, included individuals from industry, environmental organisations and government officials as peer reviewers of early drafts of their assessments. More radically, some have suggested that opening up expert knowledge to the scrutiny of the wider public is also warranted. While there will always be a unique function for expert scientific reviewers to play in authenticating knowledge, this need not exclude other interested and motivated citizens from being active.

These demands for more openness in science are intensified by the embedding of the internet and Web 2.0 media as central features of many people's social exchanges. It is no longer tenable to believe that warranted and trusted scientific knowledge can come into existence inside laboratories that are hermetically sealed from such demands.

A revolution in science

So we have a three-fold revolution in the demands that are placed on scientific knowledge claims as they apply to investigations such as climate change:

- To be warranted, knowledge must emerge from a respectful process in which science's own internal social norms and practices are adhered to;
- To be validated, knowledge must also be subject to the scrutiny of an extended community of citizens who have legitimate stakes in the significance of what is being claimed;
- And to be empowered for use in public deliberation and policy-making, knowledge must be fully exposed to the proliferating new communication media by which such extended peer scrutiny takes place.

The opportunity that lies at the centre of these more open practices of science is to secure the gold standard of trust. And it is public trust in climate change science that has potentially been damaged as a result of the exposure of emails between researchers at the University of East Anglia's Climatic Research Unit (CRU) and their peers elsewhere. The disclosure and content of these private exchanges is only the latest in a long line of instances that point to the need for major changes in the relationship between science and the public. By this, we mean a more concerted effort to explain and engage the public in understanding the processes and practices of science and scientists, as much as explaining the substance of their knowledge and how (un)certain it is.

How well does the public understand professional peer review, for example, or the role of a workshop, a seminar and a conference in science? Does the public understand how scientists go about resolving differences of opinion or reaching consensus about an important question when the uncertainties are large? We don't mean the 'textbook' answers to such things; all practising scientists know that they do not simply follow a rulebook to do their science, otherwise it could be done by a robot. Science is a deeply human activity, and we need to be more honest about what this entails. Rather than undermining science, it would actually allow the public to place their trust more appropriately in the various types of knowledge that scientists can offer.

What should be done?

At the very least, the publication of private CRU email correspondence should be seen as a wake-up call for scientists – and especially for climate scientists. The key lesson to be learnt is that not only must scientific knowledge about climate change be publicly owned – the IPCC does a fair job of this according to its own terms

– but that in the new century of digital communication and an active citizenry, the very practices of scientific enquiry must also be publicly owned.

Unsettling as this may be for scientists, the combination of 'post-normal science' and an internet-driven democratisation of knowledge demands a new professional and public ethos in science. And there is no better place to start this revolution than with climate science. After all, it is claimed, there is no more pressing global political challenge than this. But might this episode signify something more in the unfolding story of climate change – maybe the start of a process of re-structuring scientific knowledge?

It is possible that some areas of climate science have become sclerotic, that its scientific practices have become too partisan, that its funding – whether from private or public sectors – has compromised scientists. The tribalism that some of the emails reveal suggests a form of social organisation that is now all too familiar in some sections of business and government. Public trust in science, which was damaged in the BSE scandal 13 years ago, risks being affected by this latest episode.

A citizen's panel on climate change?

It is also possible that the institutional innovation that has been the IPCC has now largely run its course. Perhaps, through its structural tendency to politicise climate change science, it has helped to foster a more authoritarian and exclusive form of knowledge production – just at a time when a globalising and wired cosmopolitan culture is demanding of science something much more open and inclusive. The IPCC was designed by the UN in the Cold War era, before the internet and before GoogleWave. Maybe we should think about how a Citizen's Panel on Climate Change might work in today's world, as well as a less centralising series of IPCC-like expert assessments.

If there are serious ecological and social issues to be attended to because of the way the world's climates are changing – as the authors of this article believe – then scientists need to take a long hard look at how they are creating, validating and mobilising scientific knowledge about climate change. Climate science alters the way we think about humanity and its possible futures. It is not the case that the science is somehow now 'finished' and that we now should simply get on with implementing it. We have decades ahead when there will be interplay between evolving scientific knowledge with persisting uncertainty and ignorance, new ways of understanding our place in the world, and new ways of being in it. A more open and a better understood science process will mean more trusted science, and will increase the chances of both 'good science' and 'good policy'.

'Show your working' is the imperative given to scientists when preparing for publication to peers. There, it refers to techniques. Now, with the public as partner in the creation and implementation of scientific knowledge in the policy domain, the injunction has a new and enhanced meaning.

45 The science and politics of climate change

2 December 2009

A commentary written for *The Wall Street Journal*[121]

> Science never writes closed textbooks. It does not offer us a holy scripture, infallible and complete.

I am a climate scientist who worked in the Climatic Research Unit (CRU) at the University of East Anglia in the 1990s. I have been reflecting on the bigger lessons to be learned from the stolen emails, some of which were mine. One thing the episode has made clear is that it has become difficult to disentangle political arguments about climate policies from scientific arguments about the evidence for Man-made climate change and the confidence placed in predictions of future change. The quality of both political debate and scientific practice suffers as a consequence.

Surveys of public opinion on both sides of the Atlantic about Man-made climate change continue to tell us something politicians know only too well: the citizens they rule over have minds of their own. In the UK, a recent survey suggested that only 41 per cent believed humans are causing climate change, 32 per cent remained unsure and 15 per cent were convinced we aren't. Similar surveys in the US have shown a recent reduction in the number of people believing in Man-made climate change.

One reaction to this 'unreasonableness' is to get scientists to speak louder, more often, or more dramatically about climate change. Another reaction from government bodies and interest groups is to use ever-more-emotional campaigning. Thus both the UK government's recent 'bedtime stories' adverts, and Plane Stupid's Internet campaign showing polar bears falling past twin towers, have attracted widespread criticism for being too provocative and scary. These instinctive reactions fail to place the various aspects of our knowledge about

121 Hulme, M. (2009) The science and politics of climate change *The Wall Street Journal (Europe)* 2 December http://online.wsj.com/article/SB10001424052748704107104574571613215771336.html.

climate change – scientific insights, political values, cultural moods, personal beliefs – in right relationship with each other. Too often, when we think we are arguing over scientific evidence for climate change, we are in fact disagreeing about our different political preferences, ethical principles and value systems.

If we build the foundations of our climate-change policies so confidently and so single-mindedly on scientific claims about what the future holds and what therefore 'has to be done', then science will inevitably become the field on which political battles are waged. The mantra becomes: Get the science right, reduce the scientific uncertainties, compel everyone to believe it ... and we will have won. Not only is this an unrealistic view about how policy gets made, it also places much too great a burden on science, certainly on climate science with all of its struggles with complexity, contingency and uncertainty.

The events of the last few weeks, involving stolen professional correspondence between a small number of leading climate scientists – so-called Climategate – demonstrate my point. Both the theft itself and the alleged contents of some of the stolen emails reveal the strong polarisation and intense antagonism now found in some areas of climate science. Climate scientists, knowingly or not, become proxies for political battles. The consequence is that science, as a form of open and critical enquiry, deteriorates while the more appropriate forums for ideological battles are ignored.

We have also seen how this plays out in public debate. In the wake of Climategate, questions were asked on the BBC's *Question Time* last week about whether or not global warming was a scam. The absolutist claims of two of the panellists – *Daily Mail* journalist Melanie Phillips, and comedian and broadcaster Marcus Brigstocke – revealed how science ends up being portrayed as a fight between two dogmas: either the evidence for Man-made climate change is all fake, or else we are so sure we know how the planet works that we can claim to have just five or whatever years to save it. When science is invoked to support such dogmatic assertions, the essential character of scientific knowledge is lost – knowledge that results from open, always questioning, enquiry that, at best, can offer varying levels of confidence for pronouncements about how the world is, or may become.

The problem then with getting our relationship with science wrong is simple: we expect too much certainty, and hence clarity, about what should be done. Consequently, we fail to engage in honest and robust argument about our competing political visions and ethical values. Science never writes closed textbooks. It does not offer us a holy scripture, infallible and complete. This is especially the case with the science of climate, a complex system of enormous scale, at every turn influenced by human contingencies. Yes, science has clearly revealed that humans are influencing global climate and will continue to do so, but we don't know the full scale of the risks involved, nor how rapidly they will evolve, nor indeed – with clear insight – the relative roles of all the forcing agents involved at different scales.

Similarly, we endow analyses about the economics of climate change with too much scientific authority. Yes, we know there is a cascade of costs involved in mitigating, adapting to or ignoring climate change, but many of these costs are heavily influenced by ethical judgements about how we value things, now and in the future. These are judgements that science cannot prescribe.

The central battlegrounds on which we need to fight out the policy implications of climate change concern matters of risk management, of valuation, and political ideology. We must move the locus of public argumentation here not because the science has somehow been 'done' or 'is settled'; science will never be either of these things, although it can offer powerful forms of knowledge not available in other ways. It is a false hope to expect science to dispel the fog of uncertainty so that it finally becomes clear exactly what the future holds and what role humans have in causing it. This is one reason why British columnist George Monbiot wrote about Climategate, 'I have seldom felt so alone.'[122] By staking his position on 'the science', he feels alone and betrayed when some aspect of the science is undermined.

If Climategate leads to greater openness and transparency in climate science, and makes it less partisan, it will have done a good thing. It will enable science to function in the effective way it must do in public policy deliberations: not as the place where we import all of our legitimate disagreements, but one powerful way of offering insight about how the world works and the potential consequences of different policy choices. The important arguments about political beliefs and ethical values can then take place in open and free democracies, in those public spaces we have created for political argumentation.

122 Monbiot, G. (2009) Pretending the climate email leak isn't a crisis won't make it go away Posted 25 November 2009 http://www.guardian.co.uk/environment/georgemonbiot/2009/nov/25/monbiot-climate-leak-crisis-response.

46 A changing climate for the IPCC

3 February 2010

An essay written for SciDev.net[123]

The publication of false claims by the IPCC has been compounded by its imperious attitude, says professor of climate change Mike Hulme

The incorrect statement in the Fourth Assessment Report of the Intergovernmental Panel on Climate Change (IPCC) that the Himalayan glaciers could completely disappear by 2035 is remarkable in many ways. First, how could such a physically implausible claim have entered an early draft of an assessment undertaken by 'the world's leading experts', as IPCC authors are frequently described? Second, how did the claim survive several rounds of peer review from other IPCC authors and outside experts? Third, how did the claim, published in April 2007, remain unchallenged for more than two years before hitting the news headlines?

But perhaps most remarkable of all was the reaction of the IPCC chairman, Rajendra Pachauri, when the results of a specially commissioned Indian study of the glaciers challenged the IPCC's claim. He dismissed the new study as 'voodoo science'. Pachauri's haughty attitude helps explain why the controversy surrounding the mistaken claim – which, after all, is a rather minor piece of the picture of climate change impacts – is now filling newspapers, blogs and broadcast media. But to fully understand the timing of this affair we must reflect on the unexpected turn of events in the politics of climate change science over the past three months.

Under fire

The seminal moment was 'Climategate', when more than a thousand emails from the Climatic Research Unit (CRU) at the University of East Anglia, United

123 Hulme, M. (2010) A changing climate for the IPCC *SciDev.net*, 3 February http://www.scidev.net/en/opinions/a-changing-climate-for-the-ipcc-1.html.

Kingdom, were made public, either stolen or leaked. The emails made front-page news for several weeks and prompted a torrent of allegations about the conduct of some climate scientists and their attempts to withhold data. Crucially, although the leaked emails hardly constituted evidence of a global warming conspiracy, they legitimised those commentators who have challenged the scientific orthodoxy and the IPCC.

The emails gave such commentators unprecedented credibility in the eyes of the mainstream media and the public to question even more sharply, and less deferentially, the science underpinning human-induced climate change. Is the scientific evidence sound? Or have scientists been sexing up the risks and playing down the uncertainties?

The IPCC is the obvious target for such questions. It gained public status and stature through its Nobel Peace Prize, its outspoken chairman and its key role in forging consensus on the effects of climate change, and it has become the ultimate source of authority for scientific claims about climate change.

Many of its pronouncements have been used by political advocates to justify their policy prescriptions. 'As the science demands' was the cry echoing around the UN climate talks in Copenhagen in December, and, indeed, for months and years before.

Inaccurate claims

Both Climategate and the unexpected outcome of the Copenhagen talks have enabled critics to openly attack the IPCC. As a result, the false claim about the Himalayan glaciers has taken on considerable symbolic significance. It is not so much that an error was made, whether this is attributed to the original maverick scientist who made it, to the lapse in the IPCC's peer-review process, or even to Pachauri's rather arrogant defiance. No, the error's significance lies in the fact that it proves definitively that not everything written by the IPCC – or declared by its senior spokespersons – is true. So sceptics and bloggers are now scrutinising other chapters in the IPCC report as never before to find further evidence of inaccurate or poorly warranted statements and claims. And some have been found – for example, attributing the rise in disaster costs to climate change and claims that up to 40 per cent of the Amazon rainforest could react drastically to drought.

Time for change

What does all this mean? Well, it doesn't mean that the well-authenticated, headline conclusions about human impacts on the climate system are undermined. Nor does it mean that concerns about the risks of future climate change are misplaced.

But it does mean that the IPCC in its next assessment must be more scrupulous in adhering to its basic ground rules. It also probably means that the rules must be revised, especially regarding the use of non-peer-reviewed sources and the ways that reviewers' comments are handled. The danger of claiming, or being offered, ultimate authority – whether for determining how people should live or how policies should be made – is that it can leave you vulnerable to human error and poor practice.

By setting itself up as the impeccable and authoritative source of ultimate scientific knowledge about climate change, and with advocates justifying their case for action with 'as the [IPCC's] science demands', the IPCC's fall was almost inevitable.

A little less hubris from the IPCC might have made Pachauri more careful about using phrases such as 'voodoo science'. And a little less deference to science that 'demands action', and a more honest articulation of the ethical and political reasons for their proposed actions, would have left climate change campaigners in a stronger position.

47 Climategate, scientific controversy and the politics of climate change

28 June 2010

An unpublished research interview for an Imperial College, London, MSc thesis[124]

Topic 1: Climate change and society

Q Can you explain the concept raised in your book of: 'Depending on how one looks at the problem of climate change there are different solutions'?

A Well, what I do in the book is to look at the history of this idea of climate change and the notion that humans can have an influence on planetary climate and that our influence has become greater and greater over the last 100 years or so as we've industrialised and as we use much larger volumes of fossil-based energy: oil, coal and gas. And how this idea has emerged within science, but also how this idea of humans changing climate has now become a major political concern around the world, initially probably in the European and American countries, but over the last, let's say 10 or 15 years, increasingly as a global-scale problem. This idea of climate change has become something that I describe as being plastic. It has got so many different dimensions or facets to it that it can be manoeuvred by different interests to promote particular ideologies, or particular policies, or particular long-term or short-term goals.

So this idea of plasticity is something that I examine in the book. And to understand that, one has to use more than simply the tools of science. It's not just about measuring or quantifying or modelling and experimenting with the physical properties of the climate system. To understand this idea of plasticity and climate change you have to do some fairly hard work around sociology, around economics, around psychology, around religion and ethics in order to understand the many different ways in which this idea of climate change can take shape, and change shape, as it moves around the world.

---

124 This interview was conducted and recorded in the London School of Economics on 28 June 2010 and formed part of a Masters dissertation in Science Communication at Imperial College, London. The interview has been lightly edited to make it readable in printed form and section headings have been added.

Q   Right, so climate change is not just a scientific problem in the end?

A   It's very much more than this. And actually, I talk about the idea of climate change – what climate change means to people – changing just as much as the physical properties of the climate system might change. So when we talk about climate change, you can have two different dimensions. You can talk about change in the physical properties of the climate system. But you could also talk about changes in the idea, what climate change means to people in different cultures and different societies.

Q   What is the idea of climate?

A   Well, if we take a longer historical view, the very idea of climate itself has got a very interesting history that goes right the way back, certainly to Greek culture where we've got literature and artefacts, but almost certainly into pre-classical civilisations – the idea of climate and how climate in a way is a surrogate for talking about relationship between the human and nonhuman. And so climate has taken on lots of different meanings in different cultures in the past. Either it's been a sign of blessing from the gods, the divinities, or a sign of curse or judgement on wayward or immoral human behaviours. Then climate became something that could be, as it were, quantified or captured by scientific means and methods in the seventeenth and eighteenth centuries. And it became a new way of allowing societies to exploit climate resources. So the new imperialism of the Europeans in the nineteenth century for example used this idea of climate … well, the climate of South America is suitable for these types of commodities, the climates of Southeast Asia are suitable for these types of commodities. It was objectifying the idea of climate.

So, you can trace different eras and the different ways in which they have understood and used the idea of climate. And now in our contemporary world, we're very much using this idea of climate change, the fact that humans are not just a passive recipient of the climate, but are an active agent in shaping the climate. And this leads to different types of discourse when we think about the future.

Q   Can we control climate?

A   Well, I mean, what scientific inquiry had shown is that we are indeed a force of change, the fact that we emit greenhouse gases in the atmosphere, the fact that we change the cover of the land by cutting forests and by planting crops. These are activities that humans do, and they have implications for the climate system – they change the climate system. That's not a direct human control of that system, because all of these things are inadvertent, they're accidental by-products. So when we started burning coal in large quantities in the eighteenth and nineteenth centuries, we were not burning coal with the view to try to change the climate. It's an inadvertent consequence. So there's a difference between humans as an agent of change and humans being a deliberate controller of a system. We're not actually engineers in the sense that we intervene in the planet in particular ways in order to engineer a particular outcome.

This new idea that actually we *can* become engineers deliberately intruding into the atmosphere, into the skies, by injecting particles … the idea that we *can* engineer the planetary system for a particular outcome I think is an illusion or a delusion. We don't understand the intricate behaviours of the planet anything like well enough to claim that we are engineering the system. We're influencing it. We're agents of change, but I wouldn't like to suggest that we are engineering it, or that we have the knowledge or the wisdom to be able to engineer it for a particular outcome [see Chapter 35].

Q So if anyone could control 'the climate system', then they would be extremely powerful … ?

A Though as with all the engineering systems, you know, who are the people who actually are controlling the system? I mean, people talk about the metaphor of creating a thermostat for the planet. But if you are able to put aerosols into the stratosphere then by putting some more in you can cool it, but removing them you can warm it. So we could create this thermostat. But as with all thermostats the question is who controls the thermostat? Whose temperature is being regulated for? And this brings you immediately back into the questions of power, of justice and of democracy.

Topic 2: Climategate

Q Can you comment on the title Climategate and explain what happened?

A Well, the terminology that was used and heard first … the use of that terminology Climategate occurred within the 7 to 10 days after these emails had been made publicly available, and it's disputed as who exactly was the first person who coined it. James Delingpole, who's a columnist for *The Daily Telegraph* in Britain claims that he was the first person to use it – equally there's an American who claims that he was the first person to coin this phrase. By using the term Climategate of course it immediately connotes issues of malpractice, or secrecy, or conspiracy, or fraud as with all 'gates' going back to 'Watergate'.

But that label has stuck and has been used obviously as a short-hand to describe the incident of the emails being released and the content of what these emails had contained. So what has happened is that email correspondence between a number of fairly prominent climate scientists, over a period of 15 or 16 years, has been made publicly available. Who made those emails available is not clear. There are still possibilities that it was somebody from inside the University of East Anglia who had access to the servers where these emails were. Equally, it may have been someone from outside the institution who got access to these emails to release them. The police inquiry on that has not yet reported.[125] So there's certainly no imminent prospect of anyone being prosecuted for criminal behaviour.

But the greater significance as this unfolded in the days and weeks after was the way in which different commentators used these email chains, these correspondences, to cast question marks over some of the ways in which climate scientists had been analysing data, had been restricting access to data, had been potentially influencing the publication of certain articles. And a whole series of allegations of professional misconduct surfaced around the world. And that in turn led to these various inquiries that have been commissioned. We've had a Parliamentary inquiry and shortly we'll be having the outcome of the independent inquiry that the University of East Anglia commissioned, the Muir Russell Inquiry, that reports next week.[126]

Q Has there already been an investigation that has shown that there was malpractice?

A The House of Commons Select Committee reported a few months ago,[127] and they questioned particularly the behaviour of scientists regarding the release of information, the release of data, and wondered whether actually *not* releasing data was deemed to be a common practice in climate science. And although the scientist involved said that this was fairly common *not* to share data openly in this way for a variety of reasons, the House of Commons Committee did say that in a broader setting of scientific practice, this does seem rather anomalous, in that the normal expectation would be that scientists would make their data available for other scientists and analysts to scrutinise. So there's a big issue here about the availability and the access that people have to certain important data such as climate data that underpin some of the scientific narrative of climate change.

Q The emails: What did they say? Were they misinterpreted, contentious or reflective of proper scientific practice?

A Well, everything that we say in life can be misinterpreted. This conversation we're having today, if someone wanted to come and listen to it and interpret it in a certain way, they could of course interpret it that way. So, one has to have, first of all, an understanding of language and the way which language itself is ambiguous, it can mean different things to different people. Clearly … and I think one of the ways to think about the correspondence within these

---

125 The police investigation was officially closed on 18 July 2012 without any prosecution.
126 Russell, M., Boulton, G., Clarke, P., Eyton, D. and Norton, J. (2010) *The independent climate change e-mails review* University of East Anglia, Norwich, July.
127 House of Commons (2010) *The disclosure of climate data from the Climatic Research Unit at the University of East Anglia* Eighth report of session 2009/10, House of Commons Science and Technology Select Committee, HMSO, London.

emails is to ask the question: 'Does it display anything that is … that reveals explicit behaviour which would be deemed to be unprofessional or fraudulent?' And I think certainly, the Parliamentary Committee suggested that there was no evidence of that.

On the other hand, you look at this correspondence, you say: 'What impression does it give to an outside world about the attitudes and the beliefs of individual scientists?' So this is more about perception than about substance. And the perception, if you read some of the emails, the language of these emails, the perception that a lot of people have taken from this is that, well, these scientists seemed to have been engaged in a form of warfare with their opponents. It's sort of the language and the sense of being on the defensive and having to find all sorts of tactics and devices in order to defeat the enemy as it were. That sort of is the perception that a lot of people have taken from these emails.

And I think the interesting thing here about perception of scientists at war with other scientists is part of a much longer story about the science of climate change. The elements of controversy, the invasion of ideologies and of politics into the conduct of science, and as Judith Curry a scientist in America I think first coined … the notion of tribalism at work within science. And certainly, the perception, just from a straight reading of these emails is that there is one side of scientists, and they are fighting against another cohort of scientists … this notion of tribalism, of very strong identity. 'We are part of this tribe. They are part of the other tribe'. I think that's a very interesting perception to take of science. But I think it also says something about climate change itself which I can recognise within some of the correspondence that I've been involved and party to over 20 years ago. There is a very clear sense of 'us and them', of those who are on the side of angels, and those who are on the side of the devil.

And just to show how powerful this sense of self identity is within science, just last week in the *Proceedings of the National Academy of Sciences* [PNAS] in America, one of the leading American science journals, was a paper published by four scientists[128] who wanted to divide climate scientists into people that they categorised as 'convinced by the evidence' that humans are changing the climate, on the one hand. And another cohort of scientists that they called, 'unconvinced experts'. And they very explicitly created these two categories of scientists, the convinced and the unconvinced, and then they did various analysis on the qualifications of one group versus the qualifications of another group.

So, this too reveals this sense of tribalism if you will, that there are scientists who believe this, and there are scientists who don't believe that. And, reducing scientific inquiry and investigation into this very simple binary, right and wrong, yes or no, belief or unbelief … to me it's a very concerning practice that science can be reduced to such a simple categorisation. It's almost like the old adage,

---

128 Anderegg, W.R.L., Prall, J.W., Harold, J. and Schneider, S.H. (2010) Expert credibility in climate change *Proceedings of the National Academy of Sciences* 107(27), 12107–12109.

you've probably heard it, there are two types of people in the world, those who like to divide people into two types, and those who don't. And I think we see some of this at work in climate science.

Topic 3: Climate science and society

Q How can scientists believe and/or not believe in climate change? Can scientific data lead to different conclusions?

A Well, I think … you know, for me you have to start unpacking what are these believers – the believing scientists – believing in and what are the unbelieving scientists not believing in? Because it's not simply a case of here's a package that you've got to believe in or not. You've got to unpack the package. What actually is the package here of climate science? And if you spoke with virtually all scientists, you would get rather more nuanced judgements if you asked these sorts of questions.

If you said, 'Is there evidence that climate is changing significantly over the last hundred years?' And a vast majority of people would be entirely happy with that conclusion, that climate is changing significantly. A second set of questions then is around the cause of that change. Are these changes natural? Or are these changes caused by humans? And even there, you'll get different answers depending on whether you pose the question in the sense of: 'Is this change in climate all natural or is it all caused by humans?' Of course, nearly all scientists would say there is a combination of both natural and human factors at work. Even some of the most powerful critics of the scientific orthodoxy recognise that greenhouse gases emitted into the atmosphere trap outgoing heat from the planet, and that will lead to the warming of the planetary system. You can talk to any number or critics, and they will accept that basic physical theory.

So the issue isn't, 'Is climate change happening or is it not happening?' The question is, 'What is the extent of the human influence in the climate system? Is it a small influence, a medium influence, or a large influence?' And this is where things get much more nuanced. You can't reduce that to black and white, yes or no. And even the IPCC – the Intergovernmental Panel on Climate Change that reported three years ago – even they could only capture the knowledge of the scientific community using a very carefully hedged judgement, because they said that 'most of' the warming since the mid 20th century is very likely to have been caused by greenhouse gases. But it doesn't specify what 'most of' the warming is. More than half, okay, but that could be 51 per cent or it could be 99 per cent. There's actually a spectrum of scientific judgement at work here. Which is why trying to reduce this whole area of science into the simple binary, the believers and the unbelievers, is a gross simplification of scientific knowledge and scientific process as well, I would say.

And in a way, I am not too surprised to see this reduction into a binary working its way through the media. The media will often – must – simplify things. But it does concern me that scientists themselves are trying to reduce this complexity into a simple binary, which is why this PNAS paper published last week, I find a very unsettling paper. Here are some leading scientists who are trying to reduce the complexity of scientific inquiry, and the spectrum of scientific judgements about the human influence in the climate system, into a simple 'Either you're in this camp or you're in that camp'.

Q What are the implications of Climategate in terms of the public's perception of climate science 'fixed' and certain?

A Well, I think, that, I mean, there are number of things that follow from that. I think of two obvious ones. One is that it's an opportunity to explain more openly about the process of scientific inquiry, that the process of scientific inquiry – certainly into large complex and open systems like the planet where you can't perform very tightly controlled experiments in laboratory – these areas of science are always going to be open and ragged at the edges. There is always going to be uncertainty and controversy and contestation. That is the nature of scientific inquiry and process. You're going to have this spectrum of views within the scientific community. So I think it's an opportunity to explain that to a wider public.

But I think the other thing to draw from this is … when it comes to thinking about policy, what do we do about climate change? What do we do about the fact that humans are influencing the climate system? Because if you have this view as a number of people have, that we have to design our climate policy according to the science – so, the science will automatically translate into a set of policy actions – if that's the view that you have, then of course it places a tremendous premium on getting the science 'right' or on getting the science precise, or on getting everybody to agree about the science. Because if you win the science argument, you win the policy argument.

That is a false reading of the relationship between scientific inquiry and knowledge and policy development and implementation, it seems to me. You don't simply have the case that science dictates policy as some people would like to claim. And this is why so many of the battles and arguments around climate change are battles that occur over the science because people think we'll get the science sorted then we can make the policy decisions. That's a fallacy. We have to make policy decisions in the light of deep uncertainties about the future. And, actually, there's enough knowledge I would contend around climate change for us to have a set of vigorous arguments about what policies we should implement. It's not a case that we need more science or we need a stronger consensus or we need more certainty – because you're going to be waiting forever to achieve those things. And you'll also be hijacked by people who want to raise controversies around the science.

Actually, we know that humans are altering climate. We know that that carries risks. We know that greenhouse gases are a powerful agent of change and we also know that there are various bads that are associated with the unconstrained use of fossil energy. It's … not only does it change climate but it impairs the quality of air in cities which has health consequences. We know that it acidifies the oceans and that has risks for the marine ecosystems. We know that burning coal on open cookers in developing countries leads to possibly over a million premature deaths a year. So, we know that there are a lot of bads associated with the unfettered use of oil, coal, and gas. And we know – as we've seen in the Gulf of Mexico – that there are risks in oil drilling in deep sea water. We know there are a lot of bads associated with this.

So, what I argue is that there's enough knowledge to form the basis of policies to move our energy systems away from fossil fuel, not simply because 'we have to avoid climate catastrophe'. Because we don't know whether climate catastrophe will happen ever or when it will happen. Science will always be unclear or uncertain about that. I would contend that we have enough knowledge in order for us to drive forward policies on decarbonising and improving the quality of our air. Arguing that we need more and more scientific knowledge, more and more certain knowledge, a stronger and stronger consensus is a false reading of the relationship between climate science and climate policy.

Topic 4: The other climate controversies

Q Can you speak about the IPCC's 'glacier controversy'? What can we learn from these controversies in terms of how climate science works or should work?

A Well, first of all it is important to understand what the IPCC is. The IPCC is a panel of expert scientists brought together by the governments to the world under the banner of the United Nations in order to assess published knowledge around climate change. They don't do any research. They don't advise governments on policy. They don't invent policy measures. They're simply there to assess the knowledge, the published knowledge around climate change. The phrase that they use is that 'We are here to produce a synthesis of knowledge, an assessment of knowledge which is relevant for the policy makers, but it's not prescriptive.' It's not saying we have to do this or we have to do that.

And that is an interesting distinction which sometimes gets lost on both sides. Sometimes I think scientists think that they should have a more direct influence on the shaping of policy than was originally mandated to them. Equally, there are other commentators in society who think that the IPCC is there to determine the policy. It's not. So, that's one very important thing to get clear – the role of the IPCC.

The controversies that have emerged in the last few months around the IPCC are not coincidental with the Climategate emails nor coincidental with the politics that occurred in Copenhagen in December, in the sense that both of those two things worked in such a way as to loosen the apparent authority that IPCC scientific assessments had previously acquired and that's been epitomised in the award of the Nobel Peace Prize in 2007 to the IPCC. The IPCC assessments have gained this very high status, this great visibility around the world for producing scientific knowledge around climate change.

But the Climategate emails began to loosen their authority because people began asking the question: 'Well, are these scientists actually doing the work that we expect scientists to do or are they doing something else?' And the politics in Copenhagen meant that there was a lot of disillusionment around the policy pathway that a lot of people thought the world was pursuing. So to my mind, that then opened up this space for people who wanted to challenge the status that the IPCC had acquired. And so this particular thing that blew up originally was the Himalayan glacier claim, that was just a downright mistake. It was suggesting that the Himalayan glaciers might all have gone by 2035. And you can trace back some of the evidence for that to typographical error, because the '3' and the '0' were transposed. One of the early reports suggested perhaps they might be gone by 2305.

But the point is that that wasn't suddenly discovered in January [2010]. That had been known about several months if not a couple of years beforehand. So, why was it just in February that this was again brought back and the IPCC was challenged about it? It was because of these prior events in November and December, that gave this opportunity, this loosening up of the space. It's what some scholars in science and technology studies have coined when they look at knowledge controversies, they talk about knowledge controversies as 'moments of disturbance', moments when things that everybody had taken for granted and unquestioned suddenly become loosened up and open for challenge [see Chapter 50]. And this is what happened these recent months. And so the IPCC then became a legitimate target of criticism. Whereas six months ago, if those same critics had tried to do this, they would not have found the psychological or political support or momentum in order for their challenges to be sustained.

So various challenges emerged – the Himalayan glaciers and then we had other challenges around, whether the Amazon would be destroyed, whether African rain-fed agriculture would be wiped out, etc., etc. And this has lead to a series of repercussions, the most significant being that the United Nations has now commissioned an independent inquiry into the processes and practices of the IPCC. And that will report in August of this year. So, this is, again, a moment of disturbance that scholars talk about when a knowledge controversy dislodges and unsettles things that previously had been assumed to be certain. And it remains to be seen what the recommendations will be of this international review panel, whether they suggest that the IPCC does things in slightly differently ways to guarantee valid and well-accredited knowledge in its various assessments.

Q Did the recent climate controversies hurt the Copenhagen conference? What about the media that now speaks of a 'climate hangover'?

A Yes. Well, there certainly had been a tremendous and growing weight of expectation placed both by the policy process, the political process, and also by the civic action groups around climate change in the run up to Copenhagen. I mean, you could almost have felt during 2009 this growing body of advocacy and the rhetoric associated with that advocacy in the run up to Copenhagen – the last chance if we don't act now, the time to act, etc. And of course that creates a particular psychology of a mass movement which is ... it has some interesting characteristics to it. It's in the psychology of the prophetic pronouncements of apocalypse and doom and the end times. And we've seen these forms of mass psychology working not just in political contexts, but in other social or religious movements in the past.

And there was this, almost, this hysterical desire to grasp this moment to try to defuse these anxieties and these concerns that people have about climate change. And Copenhagen was to be the time to do it. And so of course ... I don't think the Climategate emails really had any bearing on the outcome of Copenhagen at all. I think the more significant factor at work in Copenhagen was this just tremendous mismatch between the expectation, the demand for salvation on the one hand, as opposed to the political realities of our world on the other ... And those political realities were the ones that proved to be more powerful.

And you could see the different dynamics at work, the political realities of what people talk about as a multi-polar world. It isn't simply about America; it is about the increasing power that China and India have in world geopolitics. It was about the inability of the European Union to provide any type of leadership that was able to steer China and America together. It was about the failure of the Danish presidency to lead a process of negotiating that kept the smaller developing nations on board. And the feeling was that the Danish presidency sided far too quickly with the American position and so alienated [other nations] ... Now these dynamics are simply the realities of a very complex, and divided, and self-interested world. And there's nothing very surprising about that either for that's what politics is.

But what it shows to me is that this way of trying to craft climate policy through these multilateral UN-led negotiations and getting everybody to agree to formal texts is a process that will never deliver anything very meaningful or satisfactory. And that was a conclusion I reached a few years ago. Back in the 1990s, I was a great believer in the UN multi-lateralism around climate change. I was campaigning, rather naively it seems now with retrospect, in the run up to the Kyoto Protocol in 1997. But a few years ago, about three or four years ago [see Chapter 30], I began to change my position and saw that this was not the way in which good robust climate policies were going to be crafted and Copenhagen was the culmination of that.

And so what we now see, you call it a hangover, other people – Sir David King today in the newspaper calls it the 'new pragmatism' – other people are calling for a set of mini-lateral approaches, so rather than trying to get everybody to agree, that actually we should make multiple tracks through smaller coalitions of interests. A group of scholars that I've been associated with in The Hartwell

Paper,[129] we are actually saying that we need to completely re-think what climate policy is about. It's not about trying to reach particular targets by particular dates. It's about trying to take small steps to eliminate some of these environmental and social bads that I've talked about. It's trying to eradicate black soot, for example, from indoor open cookers in developing countries, which apart from its climate consequences has a tremendous health burden on the poor. It's trying to find small steps where you can take action in smaller groups more quickly in order to eliminate some of these bads. It's not about trying to save the climate as some grand global project.

Topic 5: Current climate research and future perception of climate science

Q What are your current research interests? And how could the public better understand climate science?

A Well, my current research interests are around two major areas of climate change. One is to further the cultural understanding of climate and climate change which is back to where we started this interview – that climate and climate change are very much more than simply scientific concepts. They have cultural meanings that differ in different parts of the world. And they differ over time. And so some of my work is trying to show historically and geographically just how varied this idea of climate is to different peoples.

The other thing that I'm concerned about, and write about and talk about, is this interaction between climate science and climate policy. And we've touched on some of these issues as well – about the opportunities that knowledge controversies or science controversies give to help improve this relationship between science, society and policy. And so here it's a case of understanding that climate science is this process of continual inquiry that will always have very substantial uncertainties associated with any of the future risk predictions. In fact, as some climate scientists now are beginning to say, these uncertainties will probably get larger as we know more about the complexity of the system. It's not as though that by throwing an extra billion dollars at climate science you'll half the uncertainty. For an extra billion dollars you'll double the uncertainty!

So you've got to re-think the relationship between science and policy. You can't be waiting to make policy until the scientists are clearer in what they understand. Equally with the public, too, it's important for the pubic to realise that this isn't really about getting better scientific knowledge in order to make policy-making easier. The sort of policy questions that climate change throws up

129 Prins, G., Galiana, I., Green, C., Grundmann, R., Hulme, M., Korhola, A., Laird, F., Nordhaus, T., Pielke Jr., R., Rayner, S., Sarewitz, D., Shellenberger, M., Stehr, N. and Tezuka, H. (2010) *The Hartwell Paper: a new direction for climate policy after the crash of 2009* London School of Economics, London.

– as indeed a lot of other scientific issues throw up, for example around genetically modified crops or synthetic biology – are very, very challenging social, ethical and political questions. Simply more knowledge, more scientific knowledge in these areas, doesn't make the policy-making process any easier.

It's about trying to inform and mature the quality of public discussions and debates about these issues. And that requires some quite hard politics. It's not that we're going to be able to agree about everything. There are some very hard choices to be made about what are the priorities. And it seems to me that it's important – I see my role here – is to try to get this relationship between science, society and policy in a much better shape than it has been in the past. In the past, that shape has been distorted for various reasons that we've talked about in this interview. It's important to get that set of relationships into a better shape than it is at the moment.

And for me, in terms of my own politics, that has a very strong democratic sensibility to it. It's actually about involving people. And it's about challenging those who have got particular invested interests and putting those vested interests under scrutiny, whether they're in science or in government, or in civil society, or in the corporate world. For me, that is where these issues have to come together through an engagement and involvement in open democratic and free discussion.

48 The IPCC on trial: experimentation continues

21 July 2010

An essay written for *Environmental Research Web*, the editorial pages of the journal *Environmental Research Letters*[130]

The turbulence around climate science over recent months has been less about '*what* do we know about climate change?' and more about '*how* do we know what we know?' In other words, the controversial publication of the emails from the Climatic Research Unit (CRU) at the University of East Anglia, UK, and arguments about errors in the IPCC report have raised important questions about the *process* of scientific knowledge-making rather than seriously challenging the core *substance* of that knowledge. Here I reflect on the experimental nature of the IPCC and why that means it is essential for the IPCC to learn from past mistakes and to reflect on the changing requirements for making authoritative public knowledge in a fast-changing world.

The American scientist Roger Revelle famously wrote in the 1950s that 'human beings are now carrying out a large-scale geophysical experiment of a kind that could not have happened in the past nor be reproduced in the future'.[131] More than 50 years on we are well advanced with this experiment. Greenhouse-gas concentrations continue to rise, atmospheric aerosol loadings reflect human activities, and climates around the world are beginning to take on less familiar characters. We do not yet know the final outcome of the experiment.

Many would say that we are now engaged in another large-scale experiment, the like of which we also haven't seen before. We are in the early stages of a worldwide socio-political trial to see whether the whole panoply of human behaviours, preferences and practices can be directed towards achieving one over-arching goal: to neutralise the effects of Revelle's experiment by reducing global greenhouse-gas emissions and the production of other climate-changing agents.

130 Hulme, M. (2010) The IPCC on trial: experimentation continues *Environmental Research Web* Talking Point http://environmentalresearchweb.org/cws/article/opinion/43250. Reproduced with permission.
131 Page 19 in: Revelle, R. and Suess, H.E. (1957) Carbon dioxide exchange between atmosphere and ocean and the question of an increase of atmospheric CO_2 during the past decades *Tellus* 9, 18–27.

Unlike the inadvertent geophysical experiment, this socio-political one is both deliberate and purposeful. But as with the geophysical experiment underway, we do not know where it will lead, and we are even less sure about what its side effects will be.

The IPCC: a third experiment

But I believe there is a third experiment that it's important to identify and reflect on. This is the attempt to synthesise, globally, the body of scientific knowledge about the changing climate system and its present and future impact on matters of human concern. This experiment started, I suggest, in 1988 with the formation of the Intergovernmental Panel on Climate Change and is continuing in its most recent phase through the preparation of the Fifth IPCC Assessment Report due to be published in 2013/14. It is an undertaking that in some way connects the two other experiments: how do we organise and mobilise our limited human understanding of Revelle's geophysical experiment in such a way as to be useful for designing the emerging socio-political experiment?

Why do I suggest that the IPCC knowledge about climate change is experimental? The IPCC has three key characteristics: it is an *international* knowledge assessment; it is *multidisciplinary*; and it is governed by an *intergovernmental* process. These three characteristics distinguish the assessed knowledge produced by the IPCC from what might be called primary disciplinary knowledge generated by scientific research and scholarship. For the latter there are well-established and well-tested rules, protocols and practices that extend back to the emergence of modern science in the late seventeenth century. But the conventions and procedures that govern the international, multidisciplinary and intergovernmental process of IPCC knowledge-making have to be made up on the hoof. The IPCC is experimenting with methods of knowledge production and so brings into being a new hybrid form of scientific knowledge.

Blank canvas

When the IPCC was created in 1988 under a formal mandate from the United Nations, there was no existing model to follow, no analogous institution to copy. Decisions had to be taken from scratch on how expert authors from around the world were to be recruited, what forms of scientific knowledge were to be assessed, how successive drafts were to be checked and reviewed, and what roles governmental and non-governmental interests and expertise should play. Such an ambitious global form of public knowledge-making had not been attempted before. It was very different from making knowledge in the lab, the field or the library.

Because this has been an experimental process, the IPCC's rules and procedures have changed or been adapted over time. For example, the new role of review editor was introduced after the Second Assessment Report of 1995, and more specific guidelines have been developed on the use of non peer-reviewed, or grey, literature. All reviewer comments and the authors' responses to these comments are now made public. And formal attempts have been made to adopt a particular linguistic vocabulary to convey different levels of certainty and confidence when making specific knowledge claims, for example the phrase 'very likely' denotes a subjective probability of more than 90 per cent. The obligations of lead authors for the Fifth Assessment Report are very different – unrecognisable almost – from the quite informal duties of authors involved with the First Assessment Report in 1988.

Experimental error?

Given this context it is important to see the IPCC as producing experimental knowledge rather than yielding infallible texts (not that science ever does infallibility – only religious texts and authorities can do this). Not only does primary scientific understanding of the physical and social world evolve, but so too do the ways in which scientific knowledge is validated and made authoritative for public use. The IPCC needs to be responsive to such evolution.

So does it matter that errors, such as the incorrect date for the projected melting of Himalayan glaciers, get made? All mistakes matter, but more important is whether mistakes are honestly admitted and whether procedural lessons are learned. For an institution like the IPCC seeking to make authoritative public knowledge this learning needs to be conducted in the open, so that public confidence and trust in its judgements is retained. There are similarities here between the significance of the leaked CRU emails and the mistakes in the IPCC report. The former raised questions about how 'primary disciplinary knowledge', as I have named it here, is made; the latter episode questions how this primary knowledge is synthesised into trustworthy global hybrid knowledge. Both cases appear to have contributed to the public mood of suspicion, and in some instances the admitting of mistakes appeared to be reluctant or tardy.

As to whether lessons have been learned, a verdict is premature. The recent report from the Dutch Environmental Assessment Agency (PBL)[132] into the chapters of the IPCC Fourth Assessment on the regional impacts of climate change provides a good opportunity for learning – both about the complex nature of this hybrid IPCC knowledge and also how to improve IPCC processes and protocols for the Fifth Assessment. The PBL report identifies seven levels of IPCC assessment activity, starting with the creation of primary literature and ending with the production of the Summary for Policy-Makers (SPM) of the Synthesis Report. What becomes apparent is how heavily crafted are the core messages that emerge at the top of the IPCC pyramid. At each level, significant linguistic filtering of knowledge statements occurs before they emerge as a headline key message in the SPM of the Synthesis Report (such as 'By 2080, an increase of 5 to 8 per cent of arid and semi-arid land in Africa is projected under a range of climate

---

132 PBL (2010) *Assessing an IPCC assessment: an analysis of statements on projected regional impacts in the 2007 report* The Hague/Bilthoven, the Netherlands.

scenarios'). This is a deliberative process of collective judgement, shaped by the primary knowledge base as interpreted by individual expertise, and further modified through intergovernmental negotiation. What emerges well deserves the label of 'hybrid knowledge'.

But this is what the IPCC does and we need to understand better the experimental process by which this knowledge comes into being.

A matter of trust

What the turbulence of recent months shows is that in the making of authoritative public knowledge about climate change – knowledge that is *trustworthy*, and hence *trusted* for public use – it is the social and institutional practices of knowledge-making that matter as much as the substance of the knowledge itself. It is unrealistic to expect infallibility from scientists; but it *is* necessary to demand they be subject to some form of public accountability. Equally, it is dangerous for scientists to be insensitive to the social contexts in which their knowledge assessments are made and used; but it *is* necessary for qualified experts to be able to do what society expects experts to do – make fair and considered judgements on the basis of their expertise.

The first formal independent review of the policies and procedures of the IPCC is now underway, conducted under UN mandate by the Inter-Academy Council. A report is due after the summer,[133] for consideration by the IPCC governing body in South Korea in October. It is important that this review be fearless and thorough in its evaluation and its recommendations. In turn, the governing body of the IPCC must respond thoroughly to these recommendations. And scientists involved with the IPCC must abide by the outcomes, as part of this on-going experiment in knowledge-making.

---

133 IAC (2010) *Climate change assessments: review of the processes and procedures of the IPCC* Inter-Academy Council, Amsterdam, the Netherlands.

49 The year climate science was redefined

16 November 2010

An essay written for *The Guardian* newspaper[134]

> The 12 months since the leaking of emails written by climate change scientists have seen major shifts in environmental debate

One year ago tomorrow more than a thousand emails between scientists in the Climatic Research Unit at the University of East Anglia and their international colleagues were uploaded, unauthorised, on to a Russian FTP server. The story immediately went viral online, with lurid accusations of deception and illegality, and was soon picked up by the mainstream media. How has the climate change story changed since then? And how important was Climategate in catalysing this change? I believe there have been major shifts in how climate science is conducted, how the climate debate is framed and how climate policy is being formed. And I believe Climategate played a role in all three.

It is difficult to re-capture – or even quite believe – the cultural and political mood around climate change in the autumn of 2009. There was a rising wave of expectation that the world leaders gathering for the climate change summit in Copenhagen in December would change the world – and the climate – forever. People were fasting for climate justice, Gordon Brown was saying that Copenhagen was the last chance to reach a climate deal and there were calls for Obama to play decisively his climate card. No one 12 months ago was calling for a review of the practices of the Nobel Prize-winning Intergovernmental Panel on Climate Change, the United Nations' main climate science assessment panel. Contrarian voices, while loud, were not really being listened to. This inflated optimism had to burst and Climategate proved to be the pin.

So, 12 months later, I suggest three things of particular significance have altered. First, there has been a discernible change in some of the practices of

134 Hulme, M. (2010) The year climate science was redefined *The Guardian* 16 November http://www.guardian.co.uk/environment/2010/nov/15/year-climate-science-was-redefined.

climate science. Most obvious has been an opening up and re-analysis of some of the core observational datasets which underpin the detection of climate change trends. The UK Met Office is leading a thorough international re-analysis of 150 years of land and marine temperature data. Calls for greater transparency around scientific analysis have boosted the embryonic project of the Climate Code Foundation and its efforts to make all climate computer code open-source.

The Inter-Academy Council review[135] has recommended some significant changes in the way the IPCC assesses knowledge, in particular how it documents areas of both agreement and disagreement in the underlying science. And the Royal Society, reflecting this new mood, has issued a new guide to climate change science[136] which separates 'aspects of wide agreement', 'aspects of continuing debate' and 'aspects not well understood'. The objective of these reflexive responses in science has been to demonstrate transparency and rebuild trust.

Second, there has been a re-framing of climate change. The simple linear frame of 'here's the consensus science, now let's make climate policy' has lost out to the more ambiguous frame: 'What combination of contested political values, diverse human ideals and emergent scientific evidence can drive climate policy?' The events of the past year have finally buried the notion that scientific predictions about future climate change can be certain or precise enough to force global policy-making.

The meta-framing of climate change has therefore moved from being bi-polar – that either the scientific evidence is strong enough for action or else it is too weak for action – to being multi-polar – that narratives of climate change mobilise widely differing values which can't be homogenised through appeals to science. Those actors who have long favoured a linear connection between climate science and climate policy – spanning environmentalists, contrarians and some scientists and politicians – have been forced to rethink. It is clearer today that the battle lines around climate change have to be drawn using the language of politics, values and ethics rather than the one-dimensional language of scientific consensus or lack thereof.

Third, and perhaps most dramatically, has been the fragmentation of climate policy-making. It has been remarkable how quickly faith has evaporated in the multilateral process of the UNFCCC. Its new head, Christiana Figueres, concedes that 'there won't be a final agreement on climate change in my lifetime'. The post-mortem of COP15 showed how implausible the FAB deal wanted by NGOs – Fair, Ambitious and Binding – really was. The US Senate screwed Obama's cap-and-trade bill. And no one believes that COP16 in Cancun later this month will be any different.

135 IAC (2010) *Climate change assessments: review of the processes and procedures of the IPCC* Inter-Academy Council, Amsterdam, the Netherlands.
136 Royal Society (2010) *Climate change: a summary of the science* The Royal Society, London.

Instead, there is a new pragmatism in the air. This pragmatism has many colours and shades, but at the heart of it are three principles:

- an emphasis on the climate co-benefits of other policy innovations, such as those on health and poverty;
- a necessity to drive forward new publicly-funded investments in low-carbon energy technology;
- the cultivation of multi-level polycentric institutions and partnerships through which policy innovation may occur, rather than relying exclusively on the UN process.

These three changes are reflective of much larger cultural and political struggles regarding knowledge and power in the contemporary world which will become more salient during the next decade: the challenges to the norms of science coming from deep social and digital connectivity; the struggle to establish the appropriate cultural authority for science; and the struggles to bring democratic accountability to emergent international and global forms of governance. The shifts we are seeing around climate change are therefore symptomatic of these wider struggles.

The 12 months since 17 November 2009 have shown brutally that the social, political and cultural dynamics at work around the idea of climate change are more volatile than the slowly changing and causally entangled climate dynamics of the Earth's biogeophysical systems. Furthermore, supercomputers may mean climate science can attempt century-long predictions but that does not mean political, cultural and other unpredictable changes will not be as important.

Another IPCC assessment of scientific knowledge in four years' time is not going to make policy-making around climate change any easier. Indeed, the chances are that with scientific uncertainties and complexities about the future proliferating, and with new policy strategies such as climate geo-engineering entering the fray, further policy fragmentation around climate change is inevitable. But if such fragmentation reflects the plural, partial and provisional knowledge humans possess about the future then climate policy-making will better reflect reality. And that, I think, may be no bad thing.

50 After Climategate ... never the same

January 2013

Three years on, what have been the consequences of Climategate for climate science, for policy development and for public understandings of climate change?

Science controversy or political conspiracy?

The iconography of climate change is replete with dramatic images of the imprints of weather extremes and the effects of climate warming on physical systems. Calving icebergs are amongst the most common and images of river flooding, cyclone damage and sweltering heat are also frequently used as signifiers of climate change. These images seek to capture moments of disruption to the physical world or to the social order, thereby representing the *material* reality of climate change. But there are other moments of disruption which offer representations of the *social* reality of climate change. These could be dramatic events in the world of human discourse and political performance which also leave audiences with a sense of disorientation and disturbance. One of the most powerful of these disruptions occurred in late 2009 through the events that have commonly been dubbed 'Climategate'.

A simple account of this rupture in the discourse of climate change might read thus. Several thousand professional emails extending over a period of 13 years between a small group of influential climate scientists were 'stolen' and made public. On the basis of these emails a range of criticisms were made about the professional integrity of these scientists. In particular, had they engaged in practices which compromised key findings in climate science or, more generally, had they subverted scientific norms? A series of enquiries were conducted in the UK (and the USA) over the following nine months which largely exonerated the accused of the most serious allegations, but which also identified and criticised instances of poor communication, data management and statistical analysis. Climategate could thus be viewed as one in a rather large family of science controversies which have erupted over the years, including at the extremes Pons and Fleischmann's cold fusion, the BSE crisis and the Korean human cloning fraud.

There is, however, another way of framing Climategate. This would be to describe the publication of the emails as a criminal act by one or more individuals, motivated by a desire to undermine public trust in climate science and to deliberately discredit the status of some of the scientists involved. Rather than being a controversy about the practices of science, Climategate would then have to be placed in an even larger family of sleazy political campaigns to influence a public policy dispute, in this case whether or not to adopt stringent regulation of greenhouse gas emissions through national and international legislation.

Neither of these simplified versions of events is really adequate. To begin to understand Climategate in its broader cultural context one would need to construct a much larger story of the emerging idea of climate change over the past quarter century (see Chapter 1). The purpose of this essay, written three years after the events described above, is not to offer my own detailed account of why and how Climategate happened (I have more things to say about this, but they are for another occasion). Rather its purpose is to reflect on the repercussions of Climategate for climate science and for the public status of climate change knowledge and how citizens engage with such knowledge.[137]

As a former member[138] of the Climatic Research Unit (CRU) in the School of Environmental Sciences from where the emails had been obtained, and an employee of the University of East Anglia (UEA), I was an 'interested party' to the events unfolding in November 2009. A number of the published emails had been copied to me or else were sent to me. Some of them were sent by me. I knew a number of the central characters in the story. In the days immediately after the emails were published, I was therefore in correspondence with many interested actors in the climate change debate seeking my perspective on the significance of the unfolding events. Amongst these were scholars and scientists, members of the public, a variety of journalists and representatives of various organisations.

One of my correspondents was a senior advisor in an international NGO. In an exchange of views with him over the weekend of 5–6 December 2009, he remarked that after first learning of the emails' release two weeks earlier he had 'appreciated immediately that this would be the defining moment of the climate debate in the last five years. My initial reaction was that in terms of public perception and the "balance of legitimacy" in editorial meetings at media offices, the [CRU] emails would put us back two or three years.' In response, I offered my own view that after seeing 'the first print media story about [Climategate], I knew immediately that the climate change story will never be the same again. It's not just about a 2–3 year set-back; it is that the whole story has taken a different turn.'

---

137 In November 2010, 12 months after Climategate, I wrote a short column for *The Guardian* newspaper on this theme and which is reproduced in this book in Chapter 49. Then, I drew attention to three consequences: adjustments in scientific practices, different framings of climate change and policy fragmentation. This new essay builds and elaborates on these observations.

138 I worked in the Climatic Research Unit from 1988 to 2000, but for four years after leaving CRU I continued to use back-ups disks on one of their servers for archiving some of my files and emails. It was the security of this server that was breached in the autumn of 2009.

So from the perspective of late 2012 I am interested to explore what has been merely 'set back' two or three years by Climategate or whether, and in what ways, 'the whole story [has] taken a different turn'? Have the consequences of Climategate been good or bad; and for whom? How has climate science changed as a result? And how has the imaginative force of climate change as an idea changed over this period?

Waiting to happen

Wildfires in the American west have become more widespread and severe in recent years. Rather than finding a cause in the changing climate, ecologists have argued that forest management practices have been substantially to blame. Through modern fire suppression techniques practiced over decades, substantial amounts of dry brushwood have built up in these dryland forests which would under natural regimes have combusted through a series of small fires. When fire now is triggered through a lightning strike or a power cable spark the risk of an uncontrollable conflagration is now much greater than in the past.

I think this is a helpful analogy to explain what happened with Climategate. Over a period of 25 years or so, climate science – or at least some aspects of climate science – had slowly begun to operate in ways which was building up, so-to-speak, tinder dry brushwood. As the public policy debates around climate change grew and multiplied in scale and complexity, so the cultural and political contexts in which climate science was practiced became more febrile. If one understands science as a process sensitive to social and political context, then this inevitably began to affect how climate science was practiced. The polarising of political positions around climate change responses began to be reflected in a polarising of opinions amongst practicising climate scientists about other climate scientists: they were either 'on our side' or 'against us'. This is what Judith Curry immediately after Climategate referred to as the warring tribes of climate science.[139] So, things began to change, subtly, oh so subtly. Access to certain data became a victim of this mentality; 'friends' could access the data, but not 'enemies'. Peer review processes and judgements became exposed to similar group loyalties; papers were judged on their authorship rather than on content. The IPCC exerted an increasingly important influence over what science was deemed to be 'useful'; research was designed to fill gaps identified by the IPCC and papers were prepared to meet IPCC deadlines. And for the sake of a good cause, some climate scientists temporarily lost sight of their role to produce critical, sceptical and qualified scientific claims.

139 Such binary thinking continues to be perpetuated through analyses such as that of Anderegg et al. (2010) in which climate scientists were separated into camps 'for' and 'against' the proposition that humans are exerting a significant and growing influence on climate.

All of these practices – or variants of them – are recognisable in published ethnographies of science. In its ceaseless work of establishing and stabilising public facts, science is not immune from such influences and strategies, and never can be. But in the case of climate science in the years leading up to Climategate these influences had become magnified, as shown by Ryghaug and Skjølsvold (2010) in their analysis of the CRU emails. The CRU emails were only a shock if commentators did not realise that 'scientific facts are made and not just discovered, that they emerge as products of deliberation and persuasion, that methodological doubts may be resilient, and that scientists' trustworthiness is important' (Ryghaug and Skjølsvold, 2010: 304). And of course most did not. So these scientific 'impurities' (cf. Shapin, 2010) offered a heady mix of brushwood awaiting its lightning strike. Climategate was a wildfire waiting to happen.

Human geographer Sarah Whatmore has studied science controversies and the effects they have on science, policy and public trust and understanding. She, along with other scholars, see such events as moments of learning when 'what we think we know or, more usually, what "experts" claim to know about something [becomes] the subject of intense public interrogation' (Whatmore, 2009: 588). Controversies create opportunities for re-thinking how problems are structured, how science governs and undertakes its work and how new forms of public accountability can be exercised. From this perspective then, it is necessary to reflect on the learning that took place after Climategate. What has changed as a result of the 'intense public interrogation' which occurred during the winter of 2009/10? I am going to focus on three areas of notable impact: on scientific practice; on public opinion; and on understanding the nature of climate change scepticism.

Scientific practice

Data policy. One of the criticisms of CRU which emerged from the various inquiries into Climategate was that their data curation and data access policies were deficient. The most thorough of these reviews found 'a consistent pattern of failing to display the proper degree of openness' and that 'there was unhelpfulness is responding to requests [for data]'.[140] For these reasons, Climategate gave huge rhetorical impetus to an already emerging open-access movement, a movement which advocated the public sharing of scientific data based on two arguments: the scientific norm of communalism and the publicly funded nature of much science.

---

140 Pages 11 and 14 of Russell, M. (ed.) (2010) *The independent climate change e-mails review* University of East Anglia, Norwich. And also this comment in the Royal Society report into 'Science as an Open Enterprise': 'These [CRU] emails suggested systematic attempts to prevent access to data about one of the great global issues of the day – climate change. The researchers had failed to respond to repeated requests for sight of the data underpinning their publications, so that those seeking data had no recourse other than to use the Freedom of Information Act (FoIA) to request that the data be released' (Royal Society, 2012: 38).

These arguments were well illustrated through the events leading into Climategate: if CRU's thermometer and tree-ring data were so central to the argument that humans are influencing climate, then surely the data should be open for public scrutiny and re-analysis by any interested party? The obstructive gate-keeping role that many saw CRU adopting in this matter gave rise to suspicions of elitism and self-serving at best, and collusion and manipulation at worst.

In the last three years this drive for greater openness with scientific data has found expression in many new initiatives, both inside climate science and across science more broadly: for example journals requiring all data supporting an analysis to be accessible on-line and research funding bodies requiring all grant proposals to come with a data management and access plan. Following Climategate, CRU for the first time in its history secured external funds dedicated solely to the curation of some of its data and then in 2011 it was finally required by the UK's Information Commissioner to release all of the temperature data it had previously refused to release. As the journal *Nature Climate Change* editorialised in October 2012, 'After some false starts, and hard lessons learned, climate change researchers have woken up to the need for transparency ... and the sharing of information through public data repositories' (Anon., 2012: 703).

There are of course broader cultural and technological currents of change at work here than merely the events of Climategate; calls for 'openness' and 'transparency' now have resonance across many different social and political institutions and practices. Yet there is no doubt that Climategate functioned as a rallying point for those both inside and outside science who were arguing for much greater attention to be paid to questions of data policy and public accountability in the scientific enterprise. And it was undoubtedly one factor prompting the UK's Royal Society report on 'Science as an Open Enterprise' (Royal Society, 2012). The significance of these concerns for public trust in science was later put starkly by the Chair of the report, Geoffrey Boulton:[141]

> Science has been sleep walking into a new era ... We now have many citizens who are simply not prepared to accept the authoritative word of the scientist. They want to verify for themselves that the evidence actually justifies [the] conclusion ... [published research] conclusions [are] an opinion and unless we see the data in such a way that we can replicate it, validate it, check it, then frankly there's no reason why we should accept what they say as having any greater validity than a myth. These are not trivial issues. They are absolutely vital to the progress and delivery of science and its trustworthiness in the public domain.

Attention to uncertainty. Another of the criticisms of CRU that was upheld in the Reviews was their poor communication of uncertainty in the infamous 1999

---

141 Geoffrey Boulton, quoted in *People & Science*, Magazine of the British Association, London, September 2012, p.7

World Meteorological Organisation (WMO) graph which was designed using a data splicing 'trick' to 'hide the decline' in one of the tree-ring chronologies. One of the unsettling effects of the Climategate controversy on climate science has therefore been to encourage much more careful articulation of uncertainties concerning climate change. This is evident in communications by scientists themselves, as well as by some media commentators and reporters. After 2009 there have been more frequent reports of new scientific studies suggesting climate change may be 'less serious than previously thought', balancing the previous dominance of the tag-line 'worse than previously thought'.

Some examples of this include findings that the natural variability of stratospheric water vapour is much greater than previously thought (Solomon *et al.*, 2010), that the thermohaline circulation is more complicated than previously thought (Lozier, 2010), that soil respiration of carbon dioxide is less sensitive to temperature than assumed by climate models (Beer *et al.*, 2010), that new observations of outlet glacier velocities indicates that sea level rise from Greenland may fall well below proposed upper bounds (Moon *et al.*, 2012) and that there is no evidence for worldwide increases in drought in recent decades (Sheffield *et al.*, 2012). It is also noteworthy that in the IPCC's Special Report of Weather Extremes published in 2011 (IPCC, 2012), the language about attributing weather extreme trends to human influences was much more cautious than in the IPCC's 2007 Report.

The above are only isolated and cherry-picked examples, so what evidence is there for a more systematic adjustment in scientific practice? Using the Scopus database I searched all peer-review journal articles dealing with 'climate change' in the physical sciences for the 13 years prior to Climategate (1996–2009) and for the three years subsequent (2010–2012). The overall number of such articles continues to rise (about 6,500 in 2012 compared to about 2,000 a decade earlier), but I was interested in what proportion of these articles dealt with uncertainties. I therefore searched amongst this population for those articles which included the words 'uncertainty' or 'uncertainties' in their title, keywords or abstract. After remaining stable at around 6 per cent from 1996 to 2005, the proportion rose slightly to around 6.5 per cent between 2006 and 2008. But by 2012 the percentage had risen to 9.1.[142] If one compares the two years immediately before and after Climategate (2008–2009 with 2011–2012), then the total number of 'climate change' articles increased by about 30 per cent (from 10,047 articles to 13,111). But the number of these articles dealing with 'uncertainty' or 'uncertainties' increased by about 73 per cent (from 692 articles to 1,197). This is clear evidence

142 One finds similar *trends* in 'climate change' articles which include the term 'unknown' in their title, keywords or abstracts, although the total number of such articles is an order of magnitude fewer. And although a smaller proportion of 'climate change' articles in the non-physical sciences (life sciences, health sciences and social sciences and humanities) address 'uncertainties' (proportionally only about half as many as the physical sciences), nevertheless these too have trended the same way pre- and post-Climategate.

of a reflexive reaction by climate scientists following Climategate to engage more directly with uncertainties in their research and to communicate this in their professional publications.

Science-public dialogues. The practice of blogging dates back to the final years of the last century and large numbers of scientists now either run their own blogs or contribute to group or institutional blogs. And it was through blogs such as *Climate Audit* and *Bishop Hill*[143] that the coalescing of an on-line community of climate critics was enabled in the years leading up to Climategate, becoming an example of Jerry Ravetz's 'extended peer community' (Funtowicz and Ravetz, 1993). Many of these critical blog sites played a dominant role in shaping the early versions of the Climategate narrative.

One of the consequences of Climategate has been the increased numbers of climate scientists who are now active bloggers, either on their own bespoke sites or as visible and frequent commentators on other blogs. This trend is itself a response to my two previous observations: a new commitment from climate scientists to be open, not just about their data but even more importantly about their reasoning processes and, second, the refreshed concern with how uncertainties in climate science should be represented and interpreted in public debates.

One of the highest profile of these new bloggers is Judith Curry from Georgia Institute of Technology. She started her blog *Climate etc.*[144] in September 2010 with the aim of providing: 'a forum for climate researchers, academics and technical experts from other fields, citizen scientists, and the interested public to engage in a discussion on topics related to climate science and the science-policy interface'. Another example is from Tamsin Edwards, a climate modeller at the University of Bristol whose blog *All Models Are Wrong … but some are useful*[145] launched in January 2012 to offer 'A grown-up discussion about how to quantify uncertainties in modelling climate change and its impacts, past and future.' A third example is *Die Klimazwiebel* ('the climate onion')[146] which is unusual amongst climate change blogs for two reasons: its main bloggers are drawn from both social and natural science and it is multilingual, mainly German and English. *Die Klimazwiebel* tries to occupy a middle ground between the two warring tribes and attracts fire from both.

This trend may well have developed independently of Climategate, but it is certain that the acute controversy gave a new impetus and incentive for climate scientists to 'open-up' and explain their practices and deliberations in more public fora. The quality of some of the discussions of climate science hosted by some of these blogs has also improved, as has the range of perspectives offered. Senior

143 http://climateaudit.org and http://bishophill.squarespace.com (accessed 21 December 2012).
144 http://judithcurry.com (accessed 21 December 2012).
145 http://allmodelsarewrong.com (accessed 21 December 2012).
146 http://klimazwiebel.blogspot.co.uk/ (accessed 21 December 2012).

climate scientists, such as Richard Betts from the UK Met Office,[147] are now frequent commentators on a range of climate blogs and on-line dialogues, bringing 'institutional' climate scientists and their expertise into these new media. Important and enlightening exchanges about various aspects of climate science can now be accessed, for example about the value of the climate sensitivity on *Bishop Hill* or about the reasons for the Arctic sea-ice decline on the Dutch Met Office site *Climate Dialogue*.[148] The old in/out boundaries of climate science have been re-drawn.

The IPCC. One of the clearest repercussions of Climategate was the unprecedented challenge to the authority, accuracy and reputation of the UN's Intergovernmental Panel on Climate Change (IPCC). This was triggered early in 2010 by a story written by Fred Pearce in the 11 January issue of *New Scientist* magazine and which quickly gained global attention. This concerned a claim in the 2007 IPCC Report that the likelihood of Himalayan glaciers 'disappearing by the year 2035 and perhaps sooner is very high if the Earth keeps warming at the current rate' (IPCC, 2007: 493). Such a claim was rapidly dismissed by all experts as false and so this error rapidly prompted another investigation by the extended peer community into the veracity of other IPCC knowledge claims. A few errors and many ambiguities and poorly evidenced claims were found in the IPCC's 2007 Report, especially in its Working Group on impacts and adaptation.

Notable was the reaction in the Netherlands to an erroneously high percentage of land in that country which was claimed to lie below sea-level. Within a week the Dutch Parliament had debated the trustworthiness of the IPCC reports and voted for an independent line-by-line review of the entire Working Group 2 report. And then on 10 March, just seven weeks after the *New Scientist* article, the UN Secretary-General Ban Ki-Moon and the IPCC's parent body, the UN Environment Programme, commissioned the Inter-Academy Council (IAC) to conduct a detailed review of all of the IPCC's processes and procedures. In more than 20 years of operation, through four full Assessment Reports and winning (jointly) the award of the 2007 Nobel Peace Prize, the IPCC had never been subject to this level of scrutiny.

Yet the initial error about the Himalayan glaciers had been published by the IPCC nearly three years earlier, in April 2007, and critical journalistic attention to the claim had already been aired publicly on Indian TV in early November 2009 following a report from the Indian Government questioning the claim. The IPCC's chairman Dr Pachauri dismissed this report as 'voodoo science'. Yet this occurred ten days *before* the CRU emails were published and the criticism of the IPCC gained no traction. It was only *after* public confidence in climate science had been unsettled by Climategate that such criticism of the IPCC could 'stick'.

---

147 For Met Office scientists to be so publicly engaged would have been unimaginable even five years ago.
148 http://www.climatedialogue.org.

The repercussions of Climategate for the IPCC should therefore be seen as a good thing. The IPCC has rather helpfully been removed from its pedestal of infallibility and more plausible accounts of the knowledge-making practices of the IPCC have been established. In recent years critical scholarship has shown how the knowledge claims of the IPCC emerge from complex and contingent processes of inclusion and exclusion, where framing, personality and politics shape the resulting knowledge (e.g. O'Reilly *et al.*, 2012, on sea-level rise estimates; Mahony and Hulme, 2012, on dangerous climate change; and Suk, 2012, on climate change and malaria). Although the IAC's recommendations have not been implemented in full, its Review brought to heel an organisation and its leadership which had become high-handed, above criticism and largely unaccountable to public interests. It fully illustrated and justified the concern of scholars like Clark Miller who have drawn attention to the weak accountability of international knowledge assessment institutions.

Public opinion

One of the consequences of a public science controversy is to unsettle previously held convictions and certainties, beliefs which had been assumed but perhaps unexamined for some time. In the days immediately after the emails' release I remember a professorial colleague in the School of Environmental Sciences at UEA came to see me in my office. Knowing that I used to work in the Climatic Research Unit he wanted my candid opinion about whether our colleagues working over the bridge in CRU could indeed be trusted. Had they been manipulating data? Was the empirical evidence for global warming sound? He was being challenged to re-examine his assumed certainties; and this from someone who had worked for over 15 years in the same School as the scientists under suspicion.

This unsettling extended much more widely, although significantly it seems only to have affected certain Anglophone – UK, USA, Australia – and some northern European nations. Neighbours and friends of mine in Norwich started asking me questions about the validity of the criticisms being made. Assumed truths and certainties were being questioned. The UK environmentalist columnist George Monbiot was an example of a high profile public commentator whose beliefs were clearly challenged by the emails and subsequent allegations. 'No one has been as badly let down by the revelations in these emails as those of us who have championed the science', Monbiot wrote the week following.[149] 'I have seldom felt so alone.'

In the weeks after Climategate evidence of the impact of the controversy on public beliefs emerged from public opinion polls on both sides of the Atlantic. For

---

149 'Pretending the climate email leak isn't a crisis won't make it go away' *The Guardian* newspaper, London, 25 November 2009 http://www.guardian.co.uk/environment/george monbiot/2009/nov/25/monbiot-climate-leak-crisis-response (retrieved 26 November 2012).

example, in the USA a poll taken six weeks after the emails' release suggested that amongst those who had followed the story – just over half those surveyed – 47 per cent said it had made them more certain that 'global warming was not happening'. (A slightly larger proportion said that they had 'less trust in climate scientists' as a result). Scaled up, this amounted to about 58 million Americans who had been influenced in this way by the controversy (Maibach *et al.*, 2012).

Some have claimed that these effects on public beliefs about climate change would be relatively short-lived, but a large-scale survey in the UK conducted in March 2011 – 16 months after Climategate – suggests this may not be so (Shuckburgh *et al.*, 2012). The overall levels of concern about climate change amongst the British public had decreased over five years, almost half the population felt that the 'seriousness of climate change had been exaggerated' and one-third of the public did not trust climate scientists to tell the truth about climate change.

I don't think Climategate itself can explain all of these results and trends. Other factors such as the economy have intervened and trust across many UK public institutions and professionals has fallen, not just climate scientists. And yet what these results show is a changing and volatile public culture within which climate science is undertaken. Scientific knowledge is not created solely in the laboratory and therefore neither can it enter into public circulation simply stamped with the label 'truth'. To claim, 'I am a scientist, trust me' is no longer sufficient, even if it once was. For scientific knowledge to earn credibility as public knowledge scientists have to work as hard *outside* the laboratory as they do inside, through repeated demonstrations of their integrity, accessibility and trustworthiness. Only then will they be judged as reliable witnesses and their knowledge deemed credible (Shapin, 2010). This is not easy to do, as the events surrounding Climategate showed. What may be adequate in one culture at one moment, may not count as an adequate performance in a different context. Science is made in public as much as it is made in the laboratory or in other arcane spaces of expert deliberation.

Understanding scepticism

One of the interesting responses from the academic community since Climategate has been a new interest in studying and understanding the various manifestations of climate change scepticism.[150] One obvious reason for this interest is the evidence that voices sceptical of the standard climate change 'plan' (cf. Sarewitz,

---

150 Most of the terminology of group identities in this area is highly pejorative. Sceptics, denialists, contrarians, alarmists, heretics, believers, luke-warmists, watermelons, etc. I am using the term 'sceptic' here in the broadest way possible, wishing to reclaim a sceptical attitude as a badge of honour rather than a pathology to be overcome (as an example of the latter the American Association for the Advancement of Science used full page adverts in its journal *Science* in 2012 to offer '44 best-practices for overcoming scepticism post-Climategate').

2011) multiplied in the months following Climategate. This has been shown in the work of Painter and Ashe (2012) and Grundmann and Scott (2013) who followed media reporting of climate change around the world in the months following Climategate. Taking climate change scepticism as an object of study has engaged new scholarly communities – such as social psychologists, rhetoricians and anthropologists – and a wider range of academics than the select few sociologists who had been working in this field before. By paying attention to the political and cultural values which shape the production, circulation and reception of climate change knowledge a much richer and more helpful picture emerges. The populist notion that all climate sceptics are either in the pay of oil barons or are right-wing ideologues, as is suggested for example by studies such as Oreskes and Conway (2010), cannot be sustained.

There are many different reasons why citizens may be sceptical of aspects of climate science, certainly why they may be sceptical of knowledge claims which get exaggerated by media and lobbyists (see Chapter 38). This may be because of innate suspicion of 'big science' (which climate science has become, with powerful patrons in government and UN and international institutions) or because of a commitment to forms of data and knowledge libertarianism, as in the Wikileaks movement. Some of the individuals who pursued CRU scientists for access to data in the months leading up to Climategate may be seen in this light; they had no connections with the oil industry or conservative think-tanks. Other expressions of scepticism may result from issue fatigue, cynicism about a media who seek to sensationalise (as in the 2011 UK opinion survey quoted above) or the experience of cognitive dissonance. This latter idea captures the feeling of discomfort when someone holds two or more conflicting beliefs and Kari Marie Norgaard explores this in her ethnography of climate scepticism in a small town in Norway (Norgaard, 2011). Norgaard exposes the psychologies of climate change belief, doubt and unbelief embedded in local histories, cultures and community social practice.

But beyond these reasons for climate change scepticism, in the years following Climategate it has become more important to distinguish between at least four different aspects of the conventional climate change narrative where scepticism may emerge. Trend scepticism would be disbelieving of evidence that suggested a change in climate was occurring, whereas attribution scepticism would be doubtful that such trends were predominantly caused by human agency. Impact scepticism would question whether the melodrama of the discourse of future climate catastrophe is credible and policy scepticism would query dominant climate change policy frameworks and instruments. When this more nuanced analysis of climate change scepticism is combined with a valorisation of the scientific norm of scepticism and the democratic virtue of scrutinising and interrogating vested interests, there becomes room for more respectful arguments about what climate change signifies and what responses may be appropriate. My contention is that the events surrounding Climategate in late 2009 have opened up new spaces for such agonistic democratic virtues to be exercised.

The evolution of science

There were a number of specific circumstances and broader cultural trends which enabled the phenomenon of Climategate to erupt in November 2009 and which also shaped the competing interpretative stories in the days and weeks following. The proximate circumstances were the refusal (later deemed illegal) by CRU scientists to release climate data and the imminent COP15 climate negotiating meeting in Copenhagen. But the wider cultural trends included the growing use and visibility of social media, the Wikileaks movement, the intensification of American partisan politics and the intractability of climate change negotiations.

Scientific controversies not only reveal intellectual arguments, struggles for power and human limitations within the practices and institutions of science, they also reflect the dynamics of these exact same phenomena in the wider culture within which science takes place. And they also nearly always lead to changes in the way in which science is done as it seeks to retain its cultural authority. The nature and practice of science – how it makes authoritative knowledge about the physical world – is not defined in textbooks, least of all textbooks which are treated as timeless and universal. People have tried to define science in this way and failed. Science is like other human cultural institutions: it evolves to survive. And science controversies often become the necessary disturbances to provoke adjustment and innovation; the genetic mutations upon which processes of natural selection can operate. Whatmore observes that scientific controversies are 'generative events in their potential to foster the disordering conditions in which reasoning is forced to "slow down", creating opportunities to arouse "a different awareness of the problems and situations that mobilise us"' (Whatmore, 2009: 588).

This is certainly true of Climategate. Climate scientists, their institutions and their sponsors – i.e., climate science as an enterprise – were forced to stop and reflect on how they organised their interactions with the outside world, from data policies to language, modes of communication and forms of public engagement. The unthinking assumption that having gained broad public trust (after all the IPCC had been awarded a Nobel Prize!) this would automatically be retained, was sharply challenged. And more widely, outside science, there have been adjustments in media reporting of climate change and in the entrainment of climate science in policy deliberations, and a greater boldness from critics to challenge scientific claims and practices.

Has Climategate been a good thing? Probably not for some of the scientists caught in the conflagration. There has been some reputational damage both to individuals and institutions. The real answer though depends on one's beliefs about the nature of science and its place in public life. If one thinks of science as a pure disinterested pursuit of knowledge whose truths can then coerce social actors, whether individual or collective, into value adjustments and behavioural change, then one probably sees Climategate as a set-back. If however one understands that science only 'works' because it continually evolves norms and practices which can be rhetorically defended in public and its knowledge therefore becomes powerful *through* beliefs and behaviours, then Climategate should be seen as a creative episode. The lesson for scientists would then be this: 'In the long run, scientists may be better served by greater openness with respect to the actual practice of science, rather than upholding the conventional image of cool, restricted display of instrumental rationality' (Ryghaug and Skjølsvold, 2010: 304).

Section eight

Futures

Introduction

The first item in this section dealing with futures is a review of the IPCC's Special Report on Emissions Scenarios (SRES). My review (Chapter 51) was published in *The Times Higher Educational Supplement* early in 2001, a few months before the release of the 3rd Assessment Report of the IPCC, which relied heavily on the SRES scenarios. My review identifies the main significance of the SRES: namely, that when thinking about the future there can never be just one 'business-as-usual' scenario, but only a multiplicity of futures each of which depend upon a myriad of human choices and unforeseeable serendipities. No credible probabilities can be attached to these different scenarios. The corollary of this insight is that future climate is not determined and that human strategy and planning must account for a wide range of possible outcomes. Compare my argument here with the commentary about robust decision-making written almost a decade later (see Chapter 13).

With the publication of the IPCC 3rd Assessment in 2001, there followed a surge of new popular writings about climate change which continued to grow throughout the decade. I estimate that the rate of publication of English-language books dealing with climate change increased by a factor of at least five between 2000 and 2010. In 2004 I reviewed two of these books which offered contrasting views about climate change and the future, Mark Lynas' *High Tide* and Gus Speth's *Red Sky at Morning*. This joint review appeared in *The Times Higher Education Supplement* and is included here as Chapter 52. While Lynas takes a journalistic look at how the world was changing and may change in the future owing to climate change, Speth is more concerned with analysis of the underlying conditions and values which give rise to such rapidity of change – and how these conditions and values may change. I criticise Lynas for his travelogue style of writing and Speth for failing to forsee and engage with the shifts in the balance of geopolitical power.

Three years later – in 2007 – climate change was reaching its peak of public saliency. Ahead of the looming signs of the Global Financial Crisis (GFC), the IPCC's 4th Assessment Report had appeared, hard on the heels of the Stern Review and Al Gore's *An Inconvenient Truth* in 2006. And in October 2007, the IPCC and Al Gore were jointly awarded the Nobel Peace Prize. I took this opportunity to write a commentary for the on-line discussion forum *openDemocracy*, reflecting on the significance of this award for how climate change was being framed (Chapter 53). I found the award of the Peace Prize – of which I was one of many co-recipients – bizarre, suggesting as it did that 'good science' + 'good communication' = 'peace'. I went on to suggest that rather than solving climate change, the idea of climate change should instead be used as a magnifying glass to re-examine the multitude of diverse projects that occupy us on Earth.

Twelve months later in the autumn of 2008 the full effects of the GFC were beginning to be felt. I wrote again for *openDemocracy*, this time speculating on how the economic downturn might alter the discourse and policy of climate change (Chapter 54). I offered three scenarios: a retreat from concern about climate change; a burst of neo-Keynesian policies as in the Green New Deal; and a third option which would re-open questions about the underlying drivers of climate change and measures to redefine human well-being. I argued that the future course of climate change would be determined not by more or better science, but by the affairs of politics and culture.

The final item in this section (Chapter 55) is again a book review, this time of the Australian intellectual Clive Hamilton's book *Requiem for a Species* which had appeared earlier in 2010. Invited to review it for *Resurgence* magazine – 'at the heart of earth, art and spirit' – I took exception to the dismal pessimism being offered by Hamilton: 'Copenhagen in December 2009 was the last hope for humanity to pull back from the abyss'. As has been shown time and again, how we interpret scientific evidence is strongly conditioned by our cultural worldviews and values and Hamilton finds in the science of climate change a reflection of his own pessimistic cultural analysis. What I'm more critical of, however, is that he claims that this is the *only* reading of the scientific evidence. The human future is strongly shaped by our beliefs and the stories we tell about the future; it is not determined by the output of climate simulation models.

51 Forty ways to change a world climate

2 February 2001

A review of: *'Emissions scenarios: Special Report of the IPCC'* edited by Nebojsa Nakicenovic and Rob Swart, written for *The Times Higher Education Supplement*[151]

As the adage goes, forecasting is very difficult, especially if it is about the future. Yet concern about humanity's newly discovered and inadvertent capacity to alter the global climate forces governments, organisations and individuals to take a much longer-term view of the future than is conventional. Politicians think mostly about a single electoral cycle of four to five years; businesses for the most part are concerned with the annual profit margin or with short-term returns on investment; and we individuals are concerned mostly about events over the coming year, only occasionally raising our horizons to think perhaps about our children's education or our retirement. Decisions that we take now, and in the next few years, may well have profound effects on the climate inherited by our grandchildren and by generations beyond, and therefore greatly influence the ability of such future societies to prosper.

Current thinking suggests that the likely range of global warming during the coming century is 1.4 to 5.8°C, with a rise in average sea level of 9 to 88cm. Between about a third and a half of this range originates from the unknown future rather than from any deficiencies in our climate models. In what direction will global society move in the decades to come? Greater globalisation or Balkanisation? Stabilising of global population or doubling to 12 billion or more? Greater consumerism or a reduction in the material intensity of our lifestyles? A prolongation of our carbon-based economy or the rapid decarbonising of energy systems? It is uncertainties about these trends that underlie an important systemic source of uncertainty about future climate predictions.

This latest report from the Intergovernmental Panel on Climate Change, the *Special Report on Emissions Scenarios* (SRES), edited by Nebojsa Nakicenovic and

151 Hulme, M. (2001) *40 ways to change a world climate* Book review of 'Emissions scenarios: Special Report of the IPCC' by (eds) Nakicenovic, N. and Swart, R. in *Times Higher Education Supplement*, 2 February, p.30.

Rob Swart, matures the way we think about the future and about the climate that the future will deliver.

The IPCC is the body established in 1988 by the United Nations to provide for governments' periodic and authoritative assessments of knowledge about climate change. Previous IPCC scientific assessments have based their climate predictions on a small number of alternative greenhouse-gas emissions scenarios. In the 1990 assessment, four such scenarios were proposed – a 'business-as-usual' case and three variants. These scenarios were rather hurriedly constructed and were designed to ensure doubling of pre-industrial carbon dioxide concentrations in the atmosphere by certain fixed dates rather than being based on any coherent underlying vision of the future. The 1996 IPCC assessment relied upon six new emissions scenarios – labelled IS92a to IS92f – each of which was associated with different assumptions about future population growth, gross domestic product per capita and carbon intensity of energy supply. Although none of the six was proposed by the IPCC as normative, the *de facto* standard rapidly became the IS92a scenario around which a large majority of climate modelling and impacts, adaptation and policy analysis has been conducted in recent years.

These previous scenarios, and the way in which they were used by climate modellers, were flawed for a number of reasons. This new emissions report from the IPCC has, however, done the climate-change debate a great service in three main ways: by undermining the concept of the 'business-as-usual' scenario; by adopting for each scenario an underlying narrative vision of the future; and by adopting an open process of review and adjustment.

The SRES report clearly lays to rest the notion, still favoured by some climate modellers, of a 'business-as-usual' emissions scenario and, by association, a 'business-as-usual' climate future. As the report makes clear: 'There is no single most likely "central" or "best guess" scenario, either with respect to SRES scenarios or to the underlying literature.' The future will not be like the past, certainly not in terms of the energy, political and cultural paths that the world takes in the decades to come. To limit our thinking, and our climate modelling, to one 'preferred' future is not only the height of arrogance, it is downright dangerous. That is why Shell, and a number of other transnational corporations, routinely consider several alternative scenarios in their long-term business planning. Indeed, the SRES report benefited considerably from the intellectual input provided by one of the pioneering Shell scenario planners.

The SRES emissions scenarios are founded on storyline narrative rather than on otherwise disconnected quantitative assumptions about future population growth, GDP per capita and energy intensity. Yes, these numbers are articulated for each scenario (and comprehensive tables of these and other numbers are usefully provided in the report), but each scenario holds together because of the underlying thinking about the sort of world being described. Thus the contrast between the A1 and B1 worlds stems from the different perspectives on materialism embodied in these two scenarios. They have identical populations, and both have institutions for global governance and are technologically progressive. But the difference in values between these two worlds makes a huge difference to the demand for energy and in the uptake of different energy technologies. Consequently, the B1 world yields just over 5 billion tonnes of carbon emitted per year by 2100, compared with the A1 world with up to 30 billion tonnes – more than four times current levels. The difference for climate beyond the twenty-first century of these two scenarios, although not within the scope of the SRES report to comment on, is considerable. Of course, one can argue over whether our society will or will not reduce its material intensity in the decades to come, but that is the joy and essence of scenarios – they force us to think about the future we really want. If we do not know where we want to go, we are never going to get there.

The third laudable attribute of the SRES report has been the openness of the three-year process that fashioned the scenarios. The core writing team of 28 experts from 12 countries was supported by six modelling teams from North America, Europe and Japan. These modelling teams provided alternative qualifications of each of the narratives, a diversity that yielded an eventual set of 40 different emissions scenarios being spawned by the four core storylines. These quantifications were in turn subject to an open review process lasting nine months, whereby any research group or interested organisation was able to comment on the underlying assumptions of the quantifications themselves.

This volume is crucial contextual reading for understanding the forthcoming Third Assessment Report of the IPCC – due out in the summer – and it also provides the context for the next generation of vulnerability and adaptation studies on climate change now under way. It is essential reading for anyone who is researching in the area of future climate change and its implications. The report also has a wider relevance still in that it exemplifies an inclusive, systematic and structured approach to thinking about the global future of the twenty-first century. And in relation to climate change, it demonstrates that the future, and therefore future climate, is not given. The next time you read or hear of a climate prediction being made for the year 2100 or beyond, make sure you seek out the underlying future worldview or narrative on which that prediction has been made. The chances are it will be represented by one of the 40 scenarios in this SRES report.

The challenge to climate-change scientists is not so much to predict future climate but to provide society with the options and tools it needs to choose its own climate future. The IPCC SRES report begins to sketch out what some of these choices are, and their implications.

52 Save the world without being an eco-bore

22 October 2004

A review of: 'High Tide: News from a Warming World' by Mark Lynas and 'Red Sky at Morning: America and the Crisis of the Global Environment – A Citizen's Agenda for Action' by Gus Speth, written for *The Times Higher Education Supplement*[152]

The authors of these two books, Mark Lynas and James Gustave Speth, share a passion for safeguarding the future of this Earth; and both identify reasons why this future is threatened. They are both campaigners and these books set out, in different ways, to convince the reader of their key message: that the signs of global warming are now evident around the world (Lynas) and that the Earth's future can be sustained through citizens' actions (Speth).

On other criteria, however, the books represent very different genres of environmental journalism, from two very different people and targeted at two very different audiences. Although both are successful in communicating their key message, Lynas offers a rather superficial account that merely tugs at the emotions, while Speth's is a profound analysis that engages with mind and soul.

High Tide: News from a Warming World is a clever piece of environmental journalism that is a cross between a rather naive youthful travelogue and an apologetic for the science of climate change as assessed by the United Nation's Intergovernmental Panel on Climate Change (IPCC). Lynas starts his personal search for evidence of global warming during the autumn of 2000 in his own (flooded) backyard in the UK. He visits a flooded pub in York and a flooded housing estate in Monmouth, he talks with (also youthful) scientists and he reads the IPCC reports. But perhaps because he is not convinced by this evidence, either scientific or experiential, or maybe because he thinks that you, the reader,

152 Hulme, M. (2004) *Save the world without being an eco-bore* Book review of 'High tide: news from a warming world' by Lynas, M. and 'Red sky at morning: America and the crisis of the global environment – a citizen's agenda for action' by Speth, J.G. in *Times Higher Education Supplement*, 22 October, pp.24–25.

will not be convinced, he embarks on a worldwide trek to see if the evidence is stronger elsewhere. His odyssey takes him to Alaska, Tuvalu in the Pacific, northern China, North Carolina and Peru. In the process, he 'blows 20 years of his own personal carbon budget' on air flights and ends up at the International Climate Conference (COPE) in The Hague in November 2000, when the Bush Administration so dramatically pulled out of the process of the Kyoto Protocol.

What does Lynas conclude from his journey? Well, he is more convinced than ever that climate change is a reality and that its impacts are undeniable. He also has a wealth of stories to enable him to be 'an eco-bore at parties' and he encourages the rest of us to join him in this role. The problem is he cannot fly in a plane until 2020 – OK if he wants to travel to Rome, but not so good for his next book on the Amazon.

The idea behind this book is a good one – to paint the basic scientific conclusions of the IPCC reports about our warming world on to a more colourful and human canvas in the form of a travelogue. But somehow the book is unsatisfying for a number of reasons. The writing style is at times just too corny. The attempts to add substance and colour to what science is telling us often result in melodrama and hype, usually a sign of a weak case. Statements such as 'the world is facing climate catastrophe and we all have to change the way we live' or 'the impacts described here are just the first whispers of the hurricane of future climate change that is now bearing down on us' did not help to take my thinking to a more challenging level.

The other problem with *High Tide* is that it becomes too self-conscious and guilt-ridden. Lynas is clearly grappling with his own carbon footprint on the planet, and he never resolves this dilemma satisfactorily. If all of us were to experience Tuvalu in the way he did – 'one of those magical places of longing that can never be regained' – we would double the world's carbon emissions. Who has the right to experience such magical places and at what cost to global warming?

While one senses that Lynas is writing for a predominantly UK audience, *Red Sky at Morning* is aimed at the US market – the 'citizens' in the subtitle are certainly not, say, from Rwanda. Speth has 30 years of professional experience in environmental non-governmental organisations, he founded both the World Resources Institute and the Natural Resources Defense Council and he advised Presidents Carter and Clinton. We get a pacey, well-written account of the status of the world's environment, although the rather routine environmental litany of the first few chapters, and our inability to tackle these problems effectively, breaks no new ground.

But this book is not simply a corrective to Bjørn Lomborg's more upbeat account of the state of our environment. In the second half, Speth succeeds in engaging the mind and soul more successfully than Lynas; his years of professional experience in the campaigning world, close to the sources of power, are made to tell. Chapter five is a careful diagnosis of the failings of environmental governance worldwide and chapter six a convincing presentation of ten drivers of environmental deterioration, including the values and ethics of our (or at least US) culture. Globalisation is the focus of special attention. Speth does a good job in presenting the balance sheet of this emergent yet elusive phenomenon, distanced from the optimistic rhetoric of the Group of Eight or the compromised naivety of gap-year students.

But Speth gets even better. In chapter eight, he counters the ten drivers of deterioration with eight transitions to sustainability, including the stabilisation of population, the eradication of mass poverty (with echoes here of Lomborg's 'Copenhagen Consensus'), the support of benign technologies and the introduction of environmentally honest prices. Transitions seven and eight – concerning governance and culture and consciousness – receive particular attention. While there are better academic accounts of transition management around, for a popular readership Speth's treatment is excellent and the best starting place.

Citizenship, then, is his final rallying call and, in particular, the recognition that the deepest change of all is needed for planetary survival: a change in the way each of us sees ourselves in relation to the planet. This is about values and self-esteem, about facing our contradictory inner nature and challenging it as only reflective human beings can. The ultimate transition to sustainability has to be an inner one, from which all else flows. My only disagreement with Speth is whether it is the US that holds the key to the Earth's future. The book implicitly seems to think so, but one suspects that the twenty-first century will not only see further critical transformation of the Earth's sustaining powers but also surprising shifts in the balance of geopolitical power. Who is writing about the red skies over China, India and Russia? I hope it is not Lynas.

53 Climate change: from issue to magnifier

19 October 2007

An essay written for *openDemocracy* on-line[153]

The headline in *The Independent* newspaper on 13 October 2007 made it quite clear what the issue was: 'He's won an Oscar. He's won the Nobel Peace Prize. Now, can he win the Presidency?' Can Al Gore accomplish what no one has done before and secure this unique triumvirate of accolades and accomplishments? The award of the 2007 Nobel Peace Prize jointly to the Intergovernmental Panel on Climate Change (IPCC) (Figure 2) and the former vice-president of the United States has been applauded the world over. 'It recognises climate change as a security issue', 'it emphasises the role of science in problem-solving', 'it rewards a charismatic communicator who has put climate change centre-stage'.

Figure 2 The certificate of the award of the 2007 Nobel Peace Prize to the Intergovernmental Panel on Climate Change

153 Hulme, M. (2007) Climate change: from issue to magnifier *openDemocracy* 19 October www.opendemocracy.net/article/globalisation/politics_protest/climate_change.

To the contrary, I found the rationale for this award bizarre. It was bizarre for Alfred Nobel's peace prize to be thus awarded – I fail to see where peace has broken out as a result of climate-science papers or Al Gore presentations. And it was bizarre to join together the enterprise of a huge international scientific assessment with a one-man publicity campaign aimed at subverting the power of the White House. *The Independent* newspaper, for all its populist ballyhoo, clearly saw what it was about. Why was the Nobel committee taken in?

The limits of formula

I want to examine the thesis, this formula – implicit in the Nobel award – that good science + good communication = peace. (And here, in the context of climate change, we have to think of 'peace' as a shorthand for reducing the risks to societies posed by a warming climate). The IPCC represents good science, Al Gore and his inconvenient truth represents great communication; put them together and they can change the world. If only it were as simple as this.

This formula is very reminiscent of the deficit model of science communication, popular in the 1970s and 1980s, but now largely abandoned except in the bastions of scientistic hegemony that survive in Western liberal democracies (although sadly still more prevalent in some other parts of the world). The deficit model suggests that the reason for perverse or laggardly public policies with regard to environmental hazards is that the public and the politicians haven't grasped the science. Louder siren voices from the republic of science, crisper and more seductive communication of that science from the spin-doctors, will rectify matters. If science speaks truth to power – to use the old Quaker formulation – and speaks it persuasively through the mouth of Al Gore and the soft-lens focus of his biopic movie, then power will surely respond. Peace will break out; a runaway climate will be brought under human control.

But this is not how our world works; and this most certainly is not the way that this world is going to come to terms with its inadvertent project of climate modification. To do that, much more than just good science is needed; and it most certainly is not sufficient for that science to be filtered through the preferences and peculiarities of one man. Camilla Toulmin points out on *openDemocracy* that Gore's remedy for his climate fever, promoted in *An Inconvenient Truth* – recycle, change your lightbulb, buy a hybrid car – is not relevant for a large majority of the world's population.[154]

No, we need to understand the full significance of climate change in a different way. Yes, let us make sure that everyone understands that humans truly are altering climates around the world and that unfettered carbon-based material growth will lead to accelerated change ahead. This is what science is good at; this

154 Toulmin, C. (2007) Climate change, global justice: letter to Al Gore *openDemocracy* 12 October http://www.opendemocracy.net/globalization-climate_change_debate/letter_gore_3770.jsp.

is what good science communication should be aimed at. This is lower-case 'climate change' if you will: climate change as physical reality.

The space of difference

But at that point, we have only just started on the task required. There is also an upper-case 'Climate Change' phenomenon: Climate Change as a series of complex and constantly evolving cultural discourses. We next need to embark on the much more challenging activity of revealing and articulating the very many reasons why there is no one solution, not even one set of solutions, to (lower-case) climate change. 'Solving' climate change, 'stopping' climate chaos, 'saving' the planet in ten years are fantasy projects. We disagree about Climate Change (upper-case, its social meanings not its physical reality) not because the science is uncertain or because a few well-paid sceptics have a loud voice. We disagree about Climate Change because we disagree in quite fundamental ways about the nature of the risks posed and about what constitutes an appropriate response.

Moreover, these disagreements can be traced back to things that matter very deeply to us. They emerge from our different perceptions and tolerances of risk; from our faith in, or suspicion of, the technological genius of human engineers and innovators; from the different views we hold about the role of the state in the regulation of individual freedom; from the ways we value the natural world relative to the human world; from the beliefs we hold about the autonomy of human action relative to the idea of a divine Creator.

We have to reveal these deeper reasons why we disagree about Climate Change rather than pretending that louder, crisper and slicker communication of science will somehow bully the world to a convergence of response. In different form, but with similar intent, this has been tried before in theocracies and been found wanting. God's Ten Commandments delivered from smoke and thunder on Mount Sinai went out of fashion a while ago. *An Inconvenient Truth* is hardly an adequate substitute. As David Goldston has said with respect to the US Congress: 'the complexity of the policy discussion [about climate change] will make the previous congressional debate over whether climate change even exists seem like child's play.'[155]

This is not a prognosis for despair. It is only once we truly understand how deep our differences are, and respect them – differences in beliefs, values, goals, instruments, politics – that we will be in a position to think more clearly about what we really want to happen in the future. The role of Climate change I suggest is not a lower-case physical phenomenon to be 'solved'. Rather, we need to use the *idea* of Climate Change – the matrix of power relationships, social meanings and cultural discourses that it reveals and spawns – to rethink how we take forward our political, social and economic projects over the decades to come.

155 Goldston, D. (2007) Climate of opportunity *Nature* 445, 248 (18 January 2007).

Climate Change is a good magnifying glass for us to use in a more forensic examination than we have been used to of each of these projects – economic growth, free trade, poverty reduction, community-building, demographic management, social health, and more. Let's use the magnifying power of Climate Change – its emphasis on the long-term implications of short-term choices, its global reach, its revelation of new centres of power, its attention to both material and cultural values – to attend more closely to what we really want to achieve for humanity: affluence, justice or mere survival.

54 Amid the financial storm: re-directing climate change

1 November 2008

An essay written for *openDemocracy* on-line[156]

'Climate change' involves far more than a measured description of evolving trends in regional or global weather statistics or an uncomplicated account of the changing biogeochemical functions of the Earth system. How we talk about climate change – our discourse – is increasingly shaping our perception and interpretation of the changing physical realities that science is battling to reveal to us. At that same time, discourse is always embedded in evolving cultural, political and ethical movements and moods. Not only is our climate unstable, but how we talk about our climate is also unstable.

Understanding climate change – and the meaning we attach to the idea – is therefore always historically contingent. A sequence of four influential themes that have emerged over the last four decades illustrates the point. The idea of anthropogenic global climate change first became a possibility following the emergence of the 1960s environmental movement; the phenomenon became fully globalised during the triumph of economic globalism during the 1980s and 1990s; climate change then became part of new security discourses which emerged after 9/11; while following the Stern Report on the economics of climate change in 2006, climate change has been viewed by some as merely a consequence of market failure.

The onset of the financial crisis and a gathering worldwide recession, signalled by the banking collapses and emergency bailouts of September–October 2008, make it plausible to anticipate that this period too will generate its own characteristic frame of reference. So how will this latest manifestation of economic globalism change the way we think, talk and act about climate change? What may this new global turn eventually do to climate – both materially (by modifying the flows of carbon-dioxide through the planet) and culturally (by modifying the rhetoric and language of climate change)?

156 Hulme, M. (2008) Financial shock and climate change *openDemocracy* 1 November 2008 www.opendemocracy.net/article/amid-the-financial-storm-redirecting-climate-change.

The three pointers

Climate change is a powerful symbol of the current Zeitgeist. It is a hybrid phenomenon that reveals and is revealed by a number of important political, economic, intellectual and psychological dualisms: global-local, north-south, material-cultural, fear-hope, control-vulnerability. As we vacillate between these poles of thought and action, so too does our talk of climate change and with it our understanding of the phenomenon and what it means to us. Climate change becomes a continually mutating hybrid entity in which scientific narratives are unavoidably entangled with wider social discourses.

There are many directions in which this hybrid entity of climate change may evolve in the months and years ahead, in parallel with persisting economic problems and accompanying social dislocation. Just as the physical climate-system responds to slow-changing natural rhythms and also to more rapid human-induced perturbations, so will those human artifacts we use to make sense of climate change – language, metaphors, policies, beliefs – respond both rapidly and slowly to the new financial and economic mood. I suggest these social responses may fall into one of three meta-categories:

- a retreat from concern about climate change;
- a resurgent support for state policies on climate change;
- a deeper reflection about the human drivers of climate change.

The warm echo

The first response is seen in the instinctive reaction of some commentators to questions about the significance of the financial crisis for climate change: that the bursting of the credit-bubble will be mirrored in the bursting of the climate-change bubble. The adherents of this view will mobilise historical precedent (such as the collapse of surging environmental populism in Europe following the 1991 recession) to argue that the new recession will reveal the shallowness of the commitments of political leaders to effecting far-reaching reductions in carbon emissions.

The gradual dilution of the European Union's 2020 carbon-reduction target is reflected in the opposition of east-central European states to the auctioning of emissions-permits, and in the further transfer from Europe of 'real' physical emissions-reductions to the 'purchased' emissions-reductions of the global south; the outcome might support the contention that in this region at least, economic uncertainty is likely to be followed by environmental retreat.

The second response, and a counter-argument to the above, is that the financial crisis will provide just the confidence-boost that is needed for neo-Keynesian interventionist policies on climate change. If banks can be brought under state control through massive injection of borrowed funds, then surely new waves of state investment targeted at low-carbon energy supply and energy-efficiency technologies can be secured. This was the basic argument outlined in the 'Green New Deal', launched in July 2008 by the new economics foundation before the financial crisis fully matured; it is a case repeated by other voices such as Elliot Morley, president of GLOBE International (see his speech in Washington on 11 October 2008 to the International Commission on Climate and Energy Security).

This argument of 'crisis-as-opportunity' also resonates with the influential thinker Anthony Giddens' call for the old 'enabling state' to evolve into the new 'ensuring state'. This would require certain collectively-defined emissions-outcomes to be ensured through appropriate regulatory frameworks. To achieve the putative goals of climate policy under such a dispensation would call for, if not quite the discredited heavy-handed state collectivism of earlier ideologies, then a new 'green' authoritarianism within Western liberal democracies. In Britain for example, it is hard to see how else the climate-change minister Ed Miliband's adoption of a minimum 80 per cent reduction in carbon-emissions by 2050 will be secured. Voluntarism in climate change, this argument would suggest, may soon seem as outdated an axiom as self-regulation of capital markets now appears.

These two reactions to the financial crisis – a retreat from concern and a resurgence of regulation – appear diametrically opposed. Yet each reveals a conservative understanding of the fundamental conundrum which underlies climate change. Neither attempts or engages with any deep diagnosis of the climate-change phenomenon. A third response, less articulated at present than the other two but already visible, may by undertaking this task come to offer a new set of directions: whereby the financial crisis could move our discourse about climate change towards wider challenges to our values, our lifestyles and our (implicit and explicit) judgements about human well-being.

These challenges may range from questioning the ethics underlying institutional forms and practices to questioning the values and aspirations of individual citizens. Jonathon Porritt, for example, calls for a recognition of the parallels between financial and ecological debt; and this is itself an upstream economic version of the argument already articulated in WWFs report *Weathercocks and Signposts*.[157]

The latter suggests that no substantial progress on climate-change goals will be secured without confronting the prevailing 'extrinsic' values (material goods, financial success, image) by which society largely operates, and replacing them with 'intrinsic' values (personal growth, emotional intimacy, community involvement). This implies no less than a wholesale reframing and redirecting of human development. If we apply the scenario categories of the Intergovernmental Panel on Climate Change (IPCC's) special report on emissions scenarios, the WWF report advocates a future that looks and feels more like the 'local stewardship' (B2) scenario than the 'global affluence' (A1) world. Such a project

---

157 Crompton, T. (2008) *Weathercocks and signposts: the environment movement at a crossroads* WWF Report, Godalming, Surrey, UK.

of social restructuring resonates too with the 'low-energy cosmopolitanism' postulated by Andrew Dobson and David Hayes.

The inner cost

The financial crisis that has shaken the world in autumn 2008, and the recession trailing in its wake, will at the very least make us realise that an adequate reading of climate change is about very much more than 'getting the science right' or deploying clever Earth-systems models that offer predictions of approaching climate tipping-points. The future course of climate change – understood as a hybrid physical-cultural phenomenon in which science and our social discourse are enduringly entangled – has many more dimensions than those that can be represented numerically in a model.

The financial crisis, seen in an optimistic light, may yet do more than this. It may help us see that climate change isn't the greatest demon we humans have to confront; rather, our demons reside in the values we live by. If it causes us to expose the impossible arithmetic of more-people-plus-greater-material-consumption-plus-higher-levels-of-debt equals-an-improving-quality-of-life, then it may do more for climate change than any number of economic or political interventions.

55 A bleak analysis

November 2010

A review of: 'Requiem for a species: why we resist the truth about climate change' by Clive Hamilton, written for *Resurgence* magazine[158]

Predicting gloom and doom based solely on scientific 'facts' is missing the point, says Mike Hulme

Clive Hamilton, an Australian intellectual and former Green Party candidate, issues an impassioned call to face up to future 'realities': the world is on course for at least four degrees of global warming, and probably more; all efforts at mitigation are failing, and will fail; remediation by engineering the skies is more dangerous than the illness it is seeking to cure; and adaptation to rampant climate change will only be possible for the rich. It is time, he says, to expose the delusion that climate can be tamed; we must accept our fate.

This is a message of unrelenting pessimism and despair. Indeed, Hamilton calls upon his readers to do exactly this: to despair. 'Despair is the natural human response to the new reality we face and to resist it is to deny the truth.' This is indeed the voice of a fundamentalist eco-prophet, mimicking the hell-fire evangelical Protestant preachers of earlier eras.

Not surprisingly, there's something about this book that unsettles me. I agree with much of Hamilton's critique of modernity, which he argues lies at the root of his unfolding and unstoppable climate catastrophe, but I find that I react against his belief that runaway climate change will be the catalysing catastrophe that brings final collapse and chaos to the world. Even less do I find myself wanting to follow him into the slough of despondency to grieve in sackcloth and ashes for a lost and condemned humanity.

158 Hulme, M. (2010) Book review of *'Requeim for a species'* by Clive Hamilton *Resurgence* 263, (November/December), pp.58–59. This article is reprinted courtesy of *Resurgence & Ecologist*. To buy *Resurgence & Ecologist*, read further articles online or find out about The Resurgence Trust, visit: www.resurgence.org. All rights to this article are reserved to *Resurgence & Ecologist*, if you wish to republish or make use of this work you must contact the copyright owner to obtain permission.

And it is this paradox that troubles me. How can I explain the similarity (I think) of our underlying instinctive critique of modernity and the dehumanising effects of unbridled consumption and yet on the other hand the difference in our psychological moods; in particular, in our interpretation of the significance of the scientific claims about the future performance of the climate system?

This is not the first time I have encountered such disjunction in my public engagement with climate change. Some other academic colleagues with whom I have debated and argued these matters have echoed Hamilton's position (although perhaps not with the same brutal edge). And some reviewers and commentators have criticised my own recent book-length treatment of climate change – *Why We Disagree About Climate Change* – for being too complacent and seeming to promote passivity. 'How I learned to love climate change', they mocked.

So how do I explain this difference in perspective?

I do not doubt that human activities are now a substantial agent of change for the world's climates, and I seem to share much of Hamilton's diagnosis of the underlying ills of the social world. Yet why do we disagree so profoundly about our prognosis of what climate change signifies?

I think there are two axes along which Hamilton and I must be differently positioned: our attitudes to the veracity of scientific claims about future climate risks, and our attitudes to the resilience and never-ceasing creativity of humanity. I certainly believe Hamilton has too unquestioning a faith in the infallibility – 'truthfulness' – of science's predictive claims about future climate risks. These claims need to be separated into two different types: possible changes in physical climates due to ongoing human activities, and possible ecological and social impacts of these possible physical changes. Yet embedded in both types of claim are deep and irreducible uncertainties.

Which brings me to the second axis along which I think Hamilton and I disagree: our view of the innovative and creative potential of collective humanity. I am not here waving a magic wand and saying that we can or will innovate our way out of all the dangers the future may hold. But equally let us not denigrate ourselves. The potential creative power of humanity is now greater than it has ever been, not just in terms of the number of human brains at work, but also in how well trained they are. There are vastly more scientists, engineers, technologists and entrepreneurs alive today than there have cumulatively been in human history, and the number is growing. The pace of innovation in fields such as synthetic biology, materials science and digital communications suggests the possibility of many new and surprising technologies which make material prophecy a bold if not foolhardy profession.

Not all these inventions will be benign, of course, and of themselves they will not address the consumption fetish and spiritual malaise both Hamilton and I agree about. I share his call – who wouldn't? – for resourcefulness and selflessness to replace self-pity and instant gratification: 'values of moderation, humility and respect, even reverence for the natural world'. These are values I recognise in Christianity and in many other world religions. But I do not think it is right or effective to use an apocalyptic vision of climate change as the foundational argument for a cultural and spiritual revolution.

I believe, then, that Hamilton is placing too much weight on the foresight of science to provide his desired revolution, rather than calling for it more honestly and directly through political, psychological or spiritual engagement. (This is an argument I develop in my own book: don't use science to fight battles which are political and spiritual, especially blurry scientific claims which are easily hijacked and appropriated by one's opponents).

In the end I find Hamilton's proposed response to his picture of approaching catastrophe – Despair, Accept, Act – unconvincing. He calls on us to deliberately and consciously give up hope for the future, and embark on the cycle of grieving and reorientation. This is a big ask for an individual, let alone an entire species.

Hamilton admits he found his concluding chapter difficult to write and maybe this is why his positive vision seems rather threadbare. It seems to amount to a call for the radicalisation of democracy and civil (even violent?) disobedience against the interests of the oil and coal barons. We need 'the mobilisation of a mass movement to build a countervailing power to the elites and corporations that have captured government'.

The analysis Hamilton offers also has similarities with The Dark Mountain Project,[159] which I have seen described as 'a new cultural movement for an age of global disruption'.

Dark Mountain founders Paul Kingsnorth and Dougald Hine have been criticised (by George Monbiot among others) for giving up too soon on humanity. It is nevertheless rare to find such a gloomy, pessimistic reading of the future of humanity as is offered by Hamilton, and I am not convinced this bleak cultural analysis would be shared in some of the non-Western cultures which still see an optimistic purpose to development.

I think Hamilton is right about many of the blind and dehumanising pathways along which our modernist project is leading us. But I don't think you can reach this conclusion from the pages of IPCC reports and climate science journals in the way this author seems to be suggesting. Claiming that this is what the scientific climate forecasts really tell us is missing the point, although I fear that Hamilton would think I, too, am 'resisting the truth'.

159 http://dark-mountain.net/ (accessed 22 December 2012).

Section nine

Reactions to *Why We Disagree About Climate Change*

This essay reflects on the impact of the above book following its publication in 2009, drawing upon reactions from the academy, civic institutions, policy organisations and social commentators

January 2013

The back story

Why We Disagree About Climate Change: Understanding Controversy, Inaction and Opportunity (henceforth WWDACC) was published on 30 April 2009, seven months before the saga of Climategate and eight months before the failed Copenhagen climate negotiations. I wrote it during the winter of 2007/8 while on sabbatical leave following my resignation from the directorship of the Tyndall Centre for Climate Change Research. As I explain in the book's preface however, the structure of WWDACC was established in my mind as early as February 2003.

WWDACC was born out of my frustration that too much was being assumed as self-evidently true about climate change; for example that 'the science was settled'; that climate change was the most urgent problem facing mankind; that the only framework for climate policy was a multi-lateral targets-and-timetables approach; and that more aggressive communication of climate science through more dramatic linguistic and visual representations would convince everyone of these three truths. Resistance to such 'truths' was regarded as either irrational or else motivated by dark ideology. In a binary world there was only one side that had right on its side.

This was the dominant story of climate change I had observed develop during my time running the Tyndall Centre from 2000 to 2007. It reached a crescendo in

2006 around Al Gore's movie *An Inconvenient Truth* and his 'military missionary' movement.[160] And it didn't sit well with me. It was casually being assumed that the beliefs, values and interests which legitimised this story of climate change – what Dan Sarewitz later called 'the plan' (Sarewitz, 2011) – were self-evidently superior to any others. Or if not *this* assumption, then the argument was advanced that the risks associated with climate change were of such gravity and inevitability that legitimate differences in beliefs, values and interests should be over-ridden or suppressed because 'the science' demanded it.

Although the book developed in my mind from 2003 onwards, I didn't approach a publisher until the autumn of 2006. The first one I approached turned down my proposal without sending it out for review, commenting that 'climate change is a boom area', 'there are too many competing books', 'it would need to be globally appealing'. And the second publisher I tried was not much more enthusiastic. But serendipitous correspondence concerning another matter with the social sciences publishing director at Cambridge University Press opened the way for a more interested reaction to my proposal. Following reviews, this led eventually to a contract being signed in the summer of 2007.

On publication WWDACC gained some immediate attention – from the on-line magazine *The Register* (which adopts a rather critical stance in relation to 'the plan') and from *The Guardian* newspaper (which does not!). My interviewer at *The Register*[161] described the book in positive terms, claiming that I had 'come out of the climate change closet as an outspoken critic of such sacred cows as the UN's IPCC, the "consensus", the over-emphasis on scientific evidence in political debates about climate change, and to defend the rights of so-called "deniers" to contribute to those debates'. *The Guardian's* interviewer however was more concerned that some might take 'the message from your book, that the tone might come across as, well, let's give up on climate change. We can't do anything about it. A bit defeatist perhaps.'[162] Ahead of book reviews in academic journals, WWDACC also started attracting reviews on Amazon, including one posted in July 2009 from Joseph Bast, the CEO of the Heartland Institute in Washington DC: 'It is troubling to read a prominent scientist who has so clearly lost sight of his cardinal duty – to be skeptical of all theories and always open to new data. It is particularly troubling when this scientist endorses lying to advance his personal political agenda.'[163]

Some of the later reactions to WWDACC were therefore already foreshadowed in the spring of 2009 by these three early commentaries: I was challenging the

---

160 See *Nature* 446, 723–724 (12 April 2007): 'Al's army: Amanda Haag meets the foot soldiers of global warming'.

161 For the full interview with Stuart Blackman see Chapter 43.

162 *The Guardian's* science weekly podcast by Alok Jha, 11 May 2009, 'Why do we still disagree about climate change?' http://www.guardian.co.uk/science/audio/2009/may/11/science-weekly-podcast-climate-change. Accessed 29 November 2012.

163 http://www.amazon.com/Disagree-About-Climate-Change-Understanding/product-reviews/0521727324. Accessed 29 November 2012.

orthodox account of climate change; I was a defeatist; and I was a liar and deceiver. Then a few months later, in November 2009, the released emails from the Climatic Research Unit (CRU) provoked a major crisis in the relationship between climate scientists and some of their audiences. Monthly sales of WWDACC increased from about 400 to over 1000 during the winter of 2009/10. This boost in sales perhaps reflected people's interest in what one of the CRU emailers had to say about climate change (one commentator on the blogs even suggested that *I* had released the CRU emails to advance my book sales!). But I think this sales surge also reflected a wider public interest in why disagreements over climate change were so deep-seated and enduring, especially given the failures at COP15 in Copenhagen. WWDACC was subsequently chosen by *The Economist* magazine as one of its four science and technology 'books of the year' for 2009, it was awarded the 2010 Gerald L. Young Book Prize for human ecology and has been cited over 550 times.[164] Total sales of the book across all five continents have exceeded 15,000.

In this essay I reflect on the many reactions to WWDACC across the worlds of higher education and research, policy, business, civil society and social commentary. What has been praised, but more interestingly what has been criticised and by whom? And what impacts has the book had – both inside and outside the academy?

Reception

To understand a book it is usually important to understand something of the author. A number of reviewers of WWDACC therefore referred explicitly to my background and status as a climate scientist as being of significance. Philosopher Rupert Read stated that the great virtue of the book is that it is from 'a climate scientist [who] ... does not seek to pretend that being a climate scientist gives one any special privilege in relation to [normative] questions',[165] sociologist Steve Yearley remarked that as 'a climate modeller, a contributor to the IPCC ... he must be acutely aware of the temptation not to give an inch'[166] and geographer David Demeritt observed that I am 'steeped in the tradition of predictive environmental science' whilst offering 'a powerful critique of its limitations' (Demeritt, 2011: 133). In his Foreword to WWDACC anthropologist Steve Rayner pre-empted these observations by commenting that part of the importance of the book stemmed from this 'sociological perspective on climate change debates ... [being] ... written by a distinguished climate scientist'. Rayner explains: 'Scientists, politicians and journalists are likely to take the views of natural scientists more seriously than those of anthropologists, sociologists or political

---

164 Source: Google Scholar (6 May 2013). Equivalent citations in Scopus are 260.
165 Read, R. (2011) Review of WWDACC *Green World* 72, Spring Issue, p.23.
166 Yearley, S. (2010) Review of WWDACC *Times Higher Education*, 4 March.

scientists ... even when those scientists' own specialisations have only a tangential connection with climate science'.[167]

These observations on my credentials are interesting since by disciplinary training I am not a climate scientist but a geographer. Yes, I worked *as* a climate scientist for 12 years in the Climatic Research Unit at the University of East Anglia and I published many highly-cited papers in scientific journals during this time. But I didn't see myself writing WWDACC *as* a climate scientist and I no longer described myself as one. I wrote it from the perspective of someone who believed that understanding the arguments and discourses surrounding climate change required a wider set of disciplines, theories and perspectives than those offered merely by natural science. And yet I suspect these and other sympathetic commentators referred to my earlier scientific work to establish a certain kind of authority for the argument I advanced. If I am right about this then it is behaviour which reveals something of the theory of motivated reasoning and cognitive filtering which Dan Kahan and colleagues have explored in recent years in relation to climate change beliefs. To strengthen the receptiveness of a given audience to a certain message it is necessary to establish cultural affinity between the messenger and the audience (Kahan *et al.*, 2011). *If* arguments about climate change are widely understood primarily as arguments about science – although in WWDACC I suggest that they are not – then gaining a hearing requires an author to be accredited as 'a climate scientist'. Establishing my scientific credentials may weaken possible criticisms of my message, for example that I was writing as a social scientist without any deep understanding of climate science, that I was seeking to undermine the scientific enterprise or even that I was a climate change contrarian.

The reach of WWDACC has been extensive, both geographically and culturally. The book was reviewed formally in magazines, newspapers and academic journals around the world, in at least the following countries: Australia, Belgium, Brazil, Canada, Denmark, Germany (where a German translation is being prepared), India, Ireland, Italy, Mexico, the Netherlands, Norway, the UK and the USA. As well as newspaper broadsheet, broadcast media and internet reviews, WWDACC also gained the attention of a variety of other social networks and organisations. It was reviewed in business journals, farming, conservation and 'green' magazines, religious newsletters, policy briefings and general interest magazines. Invitations to speak – both to academic and public audiences – were received from many of the above named countries and more: Greece, Japan, New Zealand, Singapore, South Africa, Sweden, Thailand, Yemen. These invitations emanated from a broad cross-section of civil society organisations: a London-based media forum, the Australian museums sector, the international NGO Oxfam, an agri-business network, the UK Department of Energy and Climate Change, the British Council in India, science and literary festivals in Denmark,

---

167 Pages xxii–xxiii in: Rayner, S. (2009) Foreword in: *Why we disagree about climate change.*

Norway and the UK, religious organisations such as A Rocha International and individual Christian church congregations, UK high schools and colleges, the Vancouver business round-table, the Goa chamber of commerce and many more.

The above lists illustrate the reach of WWDACC around the world, but also reflect the multitude of civic and cultural interests which seek engagement with the idea of climate change. In *Why We Disagree About Climate Change* many of these interested actors found fruitful, provocative or engaging ways through which their own constituencies and audiences could connect creatively with climate change or be challenged about their responses to it.

Within the academy the reception of WWDACC can be traced rather more precisely, by studying the journals which have commissioned reviews of the book and also by following citations to the book in published literature. For example, a Scopus analysis[168] of academic journal citations to WWDACC reveals penetration across all the major meta-domains. Citations have been most numerous in the social sciences (36 per cent of citations), environmental sciences (21 per cent), earth and planetary sciences (14 per cent) and arts and humanities (8 per cent). The remaining 21 per cent of citations are within the domains of business and management, energy and engineering, medicine, psychology and agriculture. With respect to reviews in academic journals it has been the social sciences and humanities which have led the way, not surprisingly given the greater attention in these disciplines given to books. At least 26 scholarly journals have published reviews of WWDACC[169] and two of these – *The Geographical Journal* in the autumn of 2010 and *Progress in Human Geography* in winter 2011 – have published book forums, extended multiple-author analyses of the text. By engaging disciplines such as anthropology, communication science, geography, international relations, political science, religious studies, rhetoric and sociology, WWDACC goes someway to meeting Richard Norgaard's plea for a broader engagement of the academy with the challenges of environmental change. Writing in response to Bjørn Lomborg's challenge to environmental orthodoxy in *The Skeptical Environmentalist*, Norgaard pleads for academics 'to spend much more time sharing understanding across disciplines. We are operating far too distantly, we are not understanding each other, and we are not incorporating a broader understanding [of environment] into our disciplinary priorities or enquiries' (Norgaard, 2002: 292). By challenging climate change orthodoxy and through

---

168 Conducted 2 January 2013.
169 *Annals of the American Association of Geographers; Australian Journal of International Affairs; Business and Society; Climatic Change; Contemporary Sociology; Critical Policy Studies; Ecos; Energy and Environment; Futures; Gaia; Geographical Journal; Geography; Global Environmental Politics; Human Ecology Review; International Affairs; International Journal of Climate Change Strategies and Management; Journal of Integral Theory and Practice; Mexican Magazine of Foreign Affairs; Nature Climate Change; Nature and Culture; Progress in Human Geography; Public Understanding of Science; Quarterly Journal of Speech; Science; St Antony's International Review; Survival: Global Politics and Strategy.*

stimulating critical engagement with climate change amongst important disciplines which bear on the human condition, WWDACC is a response to Norgaard's call.

Criticism

A number of important criticisms of WWDACC have been made in some of the above listed reviews and in other commentaries. Let me start with the charge that it promotes a defeatist or 'quietist' attitude to climate change. This was the challenge quoted above from *The Guardian* journalist Alok Jha. At the media launch of WWDACC in May 2009 David Adam, another journalist from the same newspaper, made a more telling observation. He compared my position on climate change with the 1964 black comedy satirical film *Dr Strangelove or: How I Learned to Stop Worrying and Love the Bomb* and suggested that WWDACC was a call to stop worrying and love climate change. Other critics have made similar complaints. Philosopher Philip Kitcher in his review of WWDACC for the journal *Science* accused me of 'recommending reticence and even quietism' (Kitcher, 2010: 1230), while biologist William Anderegg laments that 'the argument of the book starts to feel uninspiring at best, paralysing at worst' (Anderegg, 2010: 659). And others have commented in public or private debates with me that the book lacks a clear rallying cry for 'decisive action' on climate change.

How valid are these criticisms? Certainly valid if, like Anderegg, one is looking for a clear road-map to agreed policy action on climate change: 'Hulme concludes without any concrete suggestions of how to move beyond the tangle of disagreement he has highlighted' (Anderegg, 2010: 659). But this frustration with WWDACC displays a belief that climate change is the sort of problem that lends itself to universal policy solutions, whether technological or social. The whole thrust of my argument in the book is that this is the wrong way to think about climate change. In one of the more perceptive reviews of WWDACC the ex-environment journalist and critic Richard D. North understood my position differently: 'Climate is not the kind of system which is readily amenable to benign influence and certainly not to the sort of control that most mainstream environmentalists, politicians, the media, and governments say they want to attempt. [Hulme] believes that attempts to benignly alter the climate could only be too feeble to make a difference, or too draconian to be politically feasible, or so large as to pose the serious risk of unintended economic, social and maybe even climate consequences'.[170] North goes on to explain that I am challenging the 'Must Do Something' orthodoxy and that this puts me at odds with 'the position of the environmentalists, mainstream political parties, and indeed everyone except the

---

170 Richard D. North's blog post, 1 December 2009, 'Climate change: let's take it seriously' http://richarddnorth.com/2009/12/climate-change-agw-lets-take-it-seriously/ (accessed 1 December 2012).

reviled sceptics or deniers ... [Hulme] believes that the proper response to [climate change] might or might not make mankind behave in ways which help the climate, or help him live with a changed climate, but they would help him live himself.'

Nevertheless, that some critics believed that I was 'giving up' on climate change and recommending a 'do-nothing' approach indicated that my call to use climate change 'to attend more closely to what we really want to achieve for ourselves and for humanity' (Hulme, 2009: 363) was either too abstract, too obscure, or both. For this and other reasons in 2010 I lent my weight behind a new initiative to re-frame climate change and climate policy which led to the publication of the Hartwell Paper (Prins *et al.*, 2010; Atkinson *et al.*, 2011). The Hartwell group of academics and policy and business analysts advocate a more pragmatic form of policy innovation, attending separately and in more de-centralised ways to matters of climate risk, atmospheric pollution and energy policy (see Chapter 34). 'Climate pragmatism' argues for environment and energy policies to be less unicentric, comprehensive and ambitious and more polycentric, targeted and modest.

Related to the above criticism is the charge that by being accommodating to a range of value-based arguments and positions on evaluating the risks of climate change I am legitimating denial: '[Hulme] thereby appears to legitimise climate change denial. This is disastrous.'[171] Both Read and Anderegg denounce my position as being too heavily influenced by 'post-modernist relativism'. What is interesting here is that a very similar criticism is made by arch-opponents of people like Read and Anderegg, namely the social commentators they would label as climate change denialists. I have already quoted from Joseph Bast of The Heartland Institute and he later satirised my invocation of JFK in his opening address to the 4th International Conference on Climate Change in May 2010, raising a laugh from his audience when he quoted me as saying, 'We need to ask not what we can do for climate change but what climate change can do for us.'[172]

But in advancing these criticisms of my 'relativism', these commentators reveal their different positions on the relationship between science and values. Read agrees with me in arguing that science is 'essentially irrelevant' to the political and evaluative judgements which lie at the heart of climate change policy (it's just that he disagrees with my normative judgements!). Anderegg and Bast on the other hand claim that science is far from irrelevant when making these judgements, but they have very different agendas for restoring science to its right and proper place in climate change debates. For Anderegg it is a case of more inter-disciplinary research and more pro-active and effective communication by climate scientists. For Bast it is a case of purifying climate science of the pernicious influences of

---

171 Read, R. op. cit.
172 The American Thinker blog, 'ICCC4 opens with a Climategate surprise', 18 May 2010, http://www.americanthinker.com/blog/2010/05/iccc_4_opens_with_a_climategat.html (accessed 1 December 2012).

'Marxists/eco-socialists' like me who subvert the norms and practices that enable scientific knowledge to be 'objective' and authoritative. These different positions lead to contradictory claims. Whereas Read claims that WWDACC strengthens the hand of neo-conservative denialists, the arch neo-conservative Bast dismisses WWDACC as pursuing a personal socialist agenda freed from the constraints of science. In a sense Read and Bast are both right, but the question about how knowledge and values, science and policy interact is more complicated than these criticisms imply.

This leads me to the first of two further criticisms which have been voiced about WWDACC: my ambiguity about where boundaries between science and policy should be drawn. In his contribution to the book review forum in *Progress in Human Geography*, David Demeritt complains that while '[Hulme] calls for us to recognise the limits of what science can and cannot tell us ... [he] is somewhat cagey about where, exactly, that line should be drawn' (Demeritt, 2011: 134). This is a criticism that has been voiced more trenchantly in unpublished correspondence with Henk van den Belt, an applied philosopher from Wageningen University, which arose following a presentation on WWDACC I made there in the autumn of 2011. Van den Belt's central problem with my position in the book – and with some of my other writings – is what he believes to be my inconsistent position about the nature of science. At times, he says, I adopt a dualist, or separatist, position seeking to purify science from the values and beliefs of its practitioners: 'We should not hide behind science when difficult ethical choices are called for ... decisions about what to do ... will always entail judgements beyond the reach of science' (Hulme, 2009: 107). At other times, he claims I seem willing to promote a co-productionist or post-normal conception of how scientific knowledge is made; i.e., that values and judgements are inseparable from the practices of science and hence from knowledge itself. He quotes from WWDACC: 'The separation of knowledge about climate change from the politics of climate change ... is no longer possible, even if it ever was' (Hulme, 2009: 107).

This criticism of ambivalence perhaps carries some weight and if I am not entirely clear about this it is because I don't think that Demeritt's clear 'lines of demarcation' can easily be drawn. Scientific knowledge is never purely objective: values are embedded throughout the making and transmission of scientific knowledge, for example from initial problem structuring to choices about communicative metaphors to representations of uncertainty. I do not subscribe to what van den Belt describes as the dualist position – the philosophical assertion that facts can be separated from values – but I would argue that when scientific knowledge is brought into political debates explicit recognition should be given to what and whose values have shaped that knowledge. Not that science reduces *to* values, but that it is a carrier *of* values.

Van den Belt's accusation of my ambivalence on this point draws attention to a lively on-going argument in the social studies of science. For example, Darrin Durant draws attention to the different positions adopted by leading science studies scholars on the question of appropriate demarcation between technical (scientific) and political (value-driven) questions (Durant, 2011). Some scholars would bracket out personal and cultural value commitments from scientific knowledge claims, clearly demarcating political from technical questions. Others would be critical of such a stance, arguing for wider social deliberation and participation across all relevant questions, thus reflecting Sheila Jasanoff's idiom of co-production between professional scientists and social interests. In my view we are left with a difficult task: the work of managing boundaries between scientific and social institutions and making clear the multiple ways in which values and interests inflect knowledge.

The remaining criticism I wish to reflect on and which I believe also carries some force is that WWDACC pays insufficient attention to 'the significance of power and political economy and the role of key actors who support and oppose climate policies in the corporate or geopolitical world of nation states' (Liverman, 2011: 135). This is a critique which is echoed by a number of reviewers. For example Max Boykoff complains similarly: 'Through more central considerations of power, Hulme could have ... interrogated how and why particular practices and ways of knowing have achieved traction, while others may have been silenced' (Boykoff, 2011: 269). And in his essay on climate change and apocalypse, Mark Levene observes in relation to WWDACC: 'We are not all equally empowered in our agreement or disagreement as to what to do (or not) about climate change. It is those with incumbent power in the richest and most powerful nations who are likely to dictate the trajectory of response' (Levene, 2010: 77).

I think these criticisms are broadly valid. Climate change is an idea that has been appropriated by many powerful actors and used to further certain interests: oil and energy corporates, the military, neo-liberal economists, deep greens, even Earth system scientists and the new planetary engineers. I did not do a good job in WWDACC at revealing and analysing the various powers at work around climate change, whether power is understood in a Gramscian or Foucauldian sense. And yet one of the arguments implicit in the book is the need to re-politicise climate change, to challenge the scientism which suggests that science should trump politics. Although not using her language, my call was to re-invigorate with respect to climate change what Chantel Mouffe describes as 'the agonistic sphere' (Mouffe, 2005). This does indeed require, as my critics demand, an exposure and analysis of blatant and latent powerful interests.

Inspiration

Notwithstanding these criticisms, WWDACC has inspired and guided the work of many. One traceable such inspiration has been the adoption and use of the book as a core text in many higher education and professional courses. It was certainly not one of my original intentions for the book to act as a student textbook and yet it has been used in senior undergraduate modules and in postgraduate and professional seminars in a surprisingly diverse number of disciplines and settings: environmental anthropology, business studies, culture and politics, development studies, English and rhetoric, environmental engineering, environmental governance, environmental studies, forestry and management, geography, history, international relations, sustainable development.[173] As Evangelos Manolas remarks in his review of WWDACC, 'There are many ways to use Hulme's book in teaching ... to make use of the many questions contained in the book ... [from] small in-class exercises ... to longer exercises such as in-class debates and take-home essay assignments.'[174]

Then there are more specific examples of how institutions and individuals for the purposes of analysing or mediating climate change discourse have made use of the framework I developed in WWDACC. The visioning project for the Australian museums sector 'Hot Science, Global Citizens' which ran between 2008 and 2011 is one such example. This project examined how museums could offer themselves as places to activate and broker discussions around climate change issues, both locally and trans-nationally. 'Hot Science, Global Citizens' drew heavily upon WWDACC in advocating the need to bring divergent cultural values into the unique public spaces offered by museums. It is in such 'safe places for dangerous ideas' where knowledge can be mediated, competing discourses and agendas tabled and debated (Cameron *et al.*, 2013).

A different type of impact of WWDACC is evidenced by ABC's TV documentary '*I can change your mind about ... climate*'.[175] First broadcast on 26 April 2012 on prime-time Australian television, the programme placed climate change critic Nick Minchin and climate change campaigner Anna Rose in dialogue with each other, as they sought out and weighed evidence from experts around the world. The programme used my framing of climate change disagreements to explore why Nick and Anna interpreted climate change evidence differently and then how these differences might be bridged in the search for common policy positions. The thesis of WWDACC that climate change can be used as a means of securing common goals in spite of antagonisms and different beliefs was supported through Nick and Anna's convergence on a policy of energy technology innovation.

One can also find many examples of how my structuring of disagreements about climate change has informed scholars interested in discourse, public policy and human values. For example Andrew Hoffman at the University of Michigan used my framing categories (see Chapter 42) in his extensive analysis of the 'logic schism' which afflicts climate change discourse in the USA (Hoffman, 2011) and

173 These are just a few of the higher education and professional institutions where WWDACC has been adopted and used as a course text: Brooklyn College, New York; CEPT University, Ahmedabad; Democratis University of Thrace; Dickinson College; London School of Economics; McGill University; NATO summer school; University of British Columbia; University of Cambridge; University College, London; University of Colorado; University of East Anglia; University of Exeter; University of Leiden; University of Manchester; University of Melbourne; University of Oregon; University of Surrey; University of Wisconsin.

174 Manalos, E. (2011) Review of WWDACC in: *International Journal of Climate Change Strategies and Management*, 3(2).

175 http://www.abc.net.au/tv/changeyourmind/ (accessed 10 December 2012).

Des Gasper of the Erasmus University in Rotterdam also used WWDACC to guide his value-critical policy analysis of climate change and international development (Gasper, 2010). A different use of WWDACC is illustrated through the work of Nick Lee and Johanna Motzkau at the University of Warwick. They used my characterisation of climate change to inform their analysis of emergent biosocial phenomena (Lee and Motzkau, 2012), new hybrid conditions of human-nature interactions which do not yield to the old models of political will and technical control. Such phenomena demand new ways of imagining human actions as always experimental, with emergent and unknowable consequences as in the case of climate change policy interventions.

The story continues

In the frontispiece to WWDACC I quote the sociologist Jonathan Haidt in saying that 'A good place to look for wisdom … is where you least expect to find it: in the minds of your opponents' and I also pay tribute to my father who showed me how disagreement can be a form of learning. So what have I learned about climate change, and maybe about myself, from the reactions to WWDACC over the last four years, some of which I have summarised here? How have the surfaced disagreements, criticisms and opposition changed the way I think about climate change?

As explained above I have been made to think more carefully about questions of power and climate change, whether conceptualised as hard, soft or diffuse power. This was a weakness in WWDACC and a lacuna in my own understanding of political science. I have also been forced to reflect more on my own position with respect to 'action' in the world. It is difficult to find evidence that WWDACC has given succour to climate change denialists and I would argue that the climate pragmatism of the Hartwell Paper is hardly defeatist. But what *are* my roles and responsibilities as an academic and as a citizen: to be a prophet, a policy advocate, a citizen activist? Is being an educator, a facilitator, an agent provocateur with respect to the many-sided phenomenon of climate change sufficient if it does not issue with a 'rallying cry' for political action? I certainly have devoted a considerable amount of time and effort to communicating and expounding my views on climate change to multiple audiences around the world, but it is not a manifesto for political action that I have been promoting. In fact my thinking on these questions has taken me to Plato, to the humanities and the philosophy of virtue ethics and a new book on these themes related to climate change will emerge soon.

I have also realised that my analysis of climate change discourse in WWDACC bisected with a number of theoretical frameworks of which I had been unaware when writing. For example until recently I had not come across the work of Chantel Mouffe and other political theorists who have diagnosed the dangers of the post-political condition. These scholars have argued that for democracy to thrive and to defuse the risks of violent conflict requires a public confrontation between clearly differentiated and articulated political visions. This is what Mouffe would describe as an agonistic public sphere protecting against antagonistic violence (Mouffe, 2005). Put in my own terms, challenging the post-political in

Western democracies requires articulating radically different responses to climate change which are rooted in different beliefs, values and interests. Only in this way is the deadening hand of a shallow cross-party political consensus on climate change – one which derives from 'the science' – going to be challenged.

Another example of my learning about empirical and theoretical work which gives substance to the rather intuitive arguments I made in WWDACC concerns the theories of biased assimilation, motivated reasoning and confirmation bias. In recent years a substantial body of work has emerged in the field of social psychology where these ideas have been empirically tested and found present in the public engagement with climate change (Kahan, 2012). My observations of the consequences of these human modes of belief formation and retention for disagreements about climate change were not as theoretically informed as they could have been. I relied heavily on Mary Douglas's and colleagues' Cultural Theory without being in a position to enrich their framework using recent social psychology scholarship.

Perhaps the overarching reflection I would make with regard to WWDACC is the contribution it has made to what may be called the cultural turn in climate change studies (see Chapter 20). This is the recognition that climate change presents far more than a challenge to climate science to make credible predictions of the climatic future, predictions which will then somehow guide present belief and behaviour. This was the thinking that motivated the IPCC into existence in 1988. Rather, what the idea of climate change has done in its contemporary form is to ask uncomfortable questions about matters such as forms of political organisation, the nature of authoritative knowledge and the human *telos*. In the end these are matters which demand cultural reflection and analysis – about human beliefs, social practices and public discourses. They are not questions that (climate) science can ever answer. And I believe WWDACC has opened up the many different ways of contemplating these more profound questions about being human and collective living. Rayner and Malone's four volume publication – *Human Choice and Climate Change* (Rayner and Malone, 1998) – led the way in this regard, but we now see a new generation of books grappling with climate change and culture: for example, climate change, media and representational practices (Boykoff, 2011; Doyle, 2011); climate change and human mobility (Hastrup and Olwig, 2012); climate change and Pacific cultures (Rudiak-Gould, 2013); the novel and climate change (Johns-Putra and Trexler (2013); and climate change and religious practice (Veldman *et al.*, 2013).

Climate change thus becomes a synecdoche – a figurative turn of phrase in which something stands in for something else – for something much more important than simply the ways humans are changing the weather. Climate change is a synecdoche for our confusion and anxiety about the goals, ambitions and destinies we foresee for ourselves and our progeny, even before we worry about whether or not we can realise them.

Bibliography

Adger, W.N., Barnett, J. and Ellemor, H. (2010) Unique and valued places at risk pp.131–140 in: *Climate change science and policy* (eds) Schneider, S., Rosencranz, A. and Mastandrea, M., Island Press, Washington DC

Agarwal, A. and Narain, S. (1991) *Global warming in an unequal world* Centre for Science and the Environment, Delhi, India

Agrawala, S. (1998) Structural and process history of the Intergovernmental Panel on Climate Change *Climatic Change* 39, 621–642

Anderegg, W.L. (2010) The ivory lighthouse: communicating climate change more effectively *Climatic Change* 101, 655–662

Anderegg, W.R.L., Prall, J.W., Harold, J. and Schneider, S.H. (2010) Expert credibility in climate change *Proceedings of the National Academy of Sciences* 107(27), 12107–12109

Anderson, K. (2005) *Predicting the weather: Victorians and the science of meteorology* Chicago University Press, Chicago IL

Anon. (1887) Professor Tyndall and the scientific movement *Nature* 36, 217–218

Anon. (1972) Quoted in: *Climatic Research Unit Monthly Bulletin* (Norwich, UEA) 1, 9

Anon. (2011) Whole system science *Nature Climate Change* 1(1), 1

Anon. (2012) Database bonanza *Nature Climate Change* 2, 703

Arrhenius, S. (1896) On the influence of carbonic acid in the air upon the temperature of the ground The *London, Edinburgh and Dublin Philosophical Magazine and Journal of Science*, 41, 237–276

Atkins, J. (2010) *Climate change for football fans: a matter of life and death* UIT Cambridge, UK

Atkinson, R., Chhetri, N., Freed, J., Galiana, I., Green, C., Hayward, S., Jenkins, J., Malone, E., Nordhaus, T., Pielke Jr., R., Prins, G., Rayner, S., Sarewitz, D. and Shellenberger, M. (2011) *Climate pragmatism: innovation, resilience and no regrets – the Hartwell analysis in an American context* The Breakthrough Institute, Washington DC

Ausubel, J. (1995) Technical progress and climate change *Energy Policy* 23, 411–416

Bardach, J.E. (1987) Global warming and the coastal zone: some effects on sites and activities Paper presented at the conference on 'Developing Policies for Responding to Future Climatic Changes', Villach, Austria, 28 September–2 October

Barnett, T.P. (1984) The estimation of 'global' sea level change: a problem of uniqueness *Journal of Geophysical Research* 89, 7980–7988

Beck, R.A. (1993) Climate, liberalism and intolerance *Weather* 48, 63–64

Beer, C. *et al.* (2010) Terrestrial gross carbon dioxide uptake: global distribution and covariation with climate *Science* 329, 834–838

Behringer, W. (2010) *A cultural history of climate* [translated from German] Polity Press, Cambridge, UK

Berkes, F., Colding, J. and Folke, C. (2000) Rediscovery of traditional ecological knowledge as adaptive management *Ecological Applications* 10(5), 1251–1262

Biermann, F. and Boas, I. (2008) Protecting climate refugees: the case for a global protocol *Environment* 50(6), 8–16

Bjurström, A. and Polk, M. (2011) Physical and economic bias in climate change research: a scientometric study of IPCC Third Assessment Report *Climatic Change* 108(1–2), 1–22

Boia, L. (2005) *The weather in the imagination* Reaktion Books, London, UK

Boykoff, M.T. (2010) Forum review *The Geographical Journal* 176(3), 267–269

Boykoff, M.T. (2011) *Who speaks for the climate? Making sense of media reporting on climate change* Cambridge University Press, Cambridge, UK

Boykoff, M.T. and Goodman, M.K. (2009) Conspicuous redemption? Reflections on the promises and perils of the 'celebritizaton' of climate change *Geoforum* 40(3), 395–406

Bradley, R.S., Diaz, H.F., Eischeid, J.K., Jones, P.D., Kelly, P.M. and Goodess, C.M. (1987) Precipitation fluctuations over northern hemisphere land areas since the mid-19th century *Science* 237, 171–755

Broadus, J., Milliman, J., Edwards, S., Aubrey, D. and Gable, F. (1986) Rising sea-level and damming of rivers: possible effects in Egypt and Bangladesh pp.165–189 in: *Effects of changes in stratospheric ozone and global climate: Volume 4, sea-level rise* (ed.) Titus, J., US EPA/UNEP, Washington DC

Brock, W.H., McMillan, N.D. and Mollan, R.C. (eds) (1981) *John Tyndall: essays on a natural philosopher* Royal Dublin Society, Dublin, Ireland

Brown, V.A., Harris, J.A. and Russell, J.Y. (eds) (2010) *Tackling wicked problems through the transdisciplinary imagination* Earthscan, London, UK

Cameron, F., Hodge, B. and Salazar, J.F. (2013) Representing climate change in museum space and places *WIREs Climate Change* 4(1), 9–21

Carson, R. (1962/2000) *Silent spring* Penguin Classics, London, UK

CCIRG (1991) *The potential effects of climate change in the United Kingdom* Department of the Environment, HMSO, London, UK

CCIRG (1996) *Potential impacts and adaptations of climate change in the United Kingdom* Department of the Environment, HMSO, London, UK

Cheng, Z.K. (1989) Global environmental changes and the Third World – methods of solving problems from the viewpoint of the Third World Paper prepared for the Conference of International Law and Global Environmental Change, University of Colorado Law School, Boulder CO, 1–3 February

Corbyn, Z. (2008) King urges arts to join crusade *Times Higher Education* 24 January www.timeshighereducation.co.uk/story.asp?storyCode=400269&sectioncode=26 (accessed 31 December 2012)

Cox, K.R. (ed.) (1997) *Spaces of globalisation: re-asserting the power of the local* Guildford Press, New York NY

Crate, S.A. and Nuttall, M. (eds) (2009) *Anthropology and climate: from encounters to actions* Left Coast Press, Walnut Creek CA

Cruikshank, J. (2005) *Do glaciers listen? Local knowledge, colonial encounters and social imagination* University of British Columbia Press, Vancouver, Canada

Crutzen, P. (2006) Albedo enhancement by stratospheric sulfur injections: a contribution to resolve a policy dilemma *Climatic Change* 77(3/4), 211–220

Dalby, S. (2007) Anthropocene geopolitics: globalisation, empire, environment and critique *Geography Compass* 1(1), 103–118

Dankelman, I. (ed.) (2010) *Gender and climate change: an introduction* Earthscan, London, UK

Davey, N. (2011) Philosophy and the quest for the unpredictable pp.303–312 in, *The public value of the humanities* (ed.) Bate.J., Bloomsbury Academic, London, UK

Demeritt, D. (2001) The construction of global warming and the politics of science *Annals of the Association of American Geographers* 91(2), 307–337

Demeritt, D. (2011) Book review symposium *Progress in Human Geography* 35(1), 132–138

Dessai, S., O'Brien, K. and Hulme, M. (2007) On uncertainty and climate change *Global Environmental Change* 17, 1–3

Diaz, H.F., Bradley, R.S. and Eischeid, J.K. (1989) Precipitation fluctuations over global land areas since the late 1800s *Journal of Geophysical Research* 94, 1195–1210

Donner, S.D. (2007) Domain of the Gods: an editorial essay *Climatic Change* 85(3–4), 231–236

Dore, M.H.I. and Guevara, R. (eds) (2000) *Sustainable forests management and global climate change* Edward Elgar, Cheltenham, UK

Doyle, J. (2011) *Mediating climate change* Ashgate Press, Farnham, UK

Durant, D. (2011) Models of democracy in social studies of science *Social Studies of Science* 41(5), 691–714

Edenhofer, O., Wallacher, J., Lotze-Campen, H., Reder, M., Knopf, B and Müller, J. (eds) (2012) *Climate change, justice and sustainability: linking climate and development policy* Springer, Dordrecht, Germany

Edwards, P.N. and Schneider, S.H. (2001) Self-governance and peer-review in science-for-policy: the case of the IPCC Second Assessment Report pp.219–246 in, *Changing the atmosphere: expert knowledge and environmental governance* Miller, C. and Edwards, P.N. (eds), MIT Press, Cambridge MA

Endfield, G. and Morris, C. (2012) Cultural spaces of climate *Climatic Change* 113(1), 1–4

Eve, A.S. and Creasey, C.H. (1945) *Life and work of John Tyndall* MacMillan & Co. Ltd, London, UK

Fagan, B.M. (2000) *The little ice age: how climate made history 1300–1850* Basic Books, New York NY

Fagan, B.M. (2004) *The long summer: how climate changed civilization* Granta Books, London, UK

Featherstone, M. and Venn, C. (2006) Problematizing global knowledge and the New Encyclopaedia Project *Theory, Culture, Society* 23, 1–20

Fine, G.A. (2007) *Authors of the storm: meteorologists and the culture of prediction* University of Chicago Press, Chicago IL

Flannery, T. (2006) *The weather makers: the history and future impact of climate change* Allen Lane, London, UK

Fleming, J.R. (1998) *Historical perspectives on climate change* Oxford University Press, Oxford, UK

Fleming, J.R. (2006) The pathological history of weather and climate modification: three cycles of promise and hype *Historical Studies in the Physical and Biological Sciences* 37(1), 3–25

Fleming, J.R. (2010) *Fixing the sky: checkered history of weather and climate control* Columbia University Press, New York NY

Fogel, C. (2004) The local, the global and the Kyoto Protocol pp.103–126 in, *Earthly politics: local and global in environmental governance* (eds) Jasanoff, S. and Martello, M.L., MIT Press, Cambridge MA

Forgan, S. (1981) Tyndall at the Royal Institution pp.49–60 in: *John Tyndall: essays on a natural philosopher* (eds) Brock, W.H., McMillan, N.D. and Mollan, R.C., Royal Dublin Society, Dublin, Ireland

Foster, P., Ramaswamy, V., Artaxo, P., Berntsen, T., Betts, R., Fahey, D.W., Jaywood, J, Lean, J., Lowe, D.C., Myhre, G., Nganga, J., Prinn, R., Raga, G., Schuz, M. and Van Dorland, R. (eds) (2007) Changes in atmospheric constituents and irradiative forcing pp.129–234 in, *Climate change 2007: the physical science basis* (eds) Solomon, S., Qin, D., Manning, M., Marquis, M., Averyt, K., Tignor, M.M.B., Miller, H.L. jr. and Chen, Z., WGI Contribution to the Fourth Assessment Report of the IPCC, Cambridge University Press, Cambridge, UK

Franz, W.E. (1997) *The development of an international agenda for climate change: connecting science to policy* Interim Report IR-97–034, IIASA, Laxenburg, Austria

Friday, J.R., MacLeod, R.M. and Shepherd, P. (1974) *John Tyndall natural philosopher 1820–1893: catalogue of correspondence, journals and collected papers* Mansell, London, UK

Fukuyama, F. (1992) *The end of history and the last man* Free Press, New York NY

Funtowicz, S.O. and Ravetz, J.R. (1993) Science for the post-normal age *Futures* 25, 739–755

Garnaut, R. (ed.) (2008) *The Garnaut climate change review* Cambridge University Press, Cambridge, UK

Gasper, D. (2010) *Influencing the climate: explorations in interpretative and value-critical policy analysis* International Institute of Social Studies, Erasmus University of Rotterdam, the Netherlands

Glacken, C. (1967) *Traces on the Rhodian Shore: nature and culture in Western thought from Ancient times to the end of the eighteenth century* University of California Press, Berkeley CA

Gleick, P.H. (1989) Climate change and international politics: problems facing developing countries *Ambio*, 18, 333–339

Golinski, J. (2007) *British weather and the climate of enlightenment* Chicago University Press, Chicago IL

Gornitz, V. (ed.) (2008) *Encyclopedia of paleoclimatology and ancient environments* Springer, New York NY

Goudie, A.S. and Cuff, D.J. (eds) (2001) *Encyclopedia of global change: environmental change and human society* Oxford University Press, New York NY

Grove, J.M. (1988) *The little ice age* Methuen, London, UK

Grove, R.H. (1994) A historical review of early institutional and conservationist responses to fears of artificially induced global climate change: the deforestation-desiccation discourse 1500–1860 *Chemosphere* 29(5), 1001–1103

Grubb, M. (1989) *The greenhouse effect: negotiating targets* Royal Institute of International Affairs, London, UK

Grundmann, R. and Scott, M. (2013) Disputed climate science in the media: do countries matter? *Public Understanding of Science* DOI: 10.1177/0963662512467732 (published on-line 11 December 2012)

Hastrup, K. and Olwig, K.F. (eds.) (2012) *Climate change and human mobility: global challenges to the social sciences* Cambridge University Press, Cambridge, UK

Hewitt, C., Mason, S. and Walland, D. (2012) The Global Framework for Climate Services *Nature Climate Change* 2(12), 831–832

Hoffman, A.J. (2011) Talking past each other? Cultural framing of sceptical and convinced logics in the climate change debate *Organisation & Environment* 24(1), 3–33

Horn, R. (2007) *Weather reports you* Artangel/Steidl, Göttingen, Germany

Howden, Daniel, Diplomats warned that climate change is security issue, not a green dilemma, *The Independent* newspaper, 6 December 2007

Huler, S. (2004) *Defining the wind: the Beaufort scale and how a nineteenth century Admiral turned science into poetry* Crown Publishers, New York

Hulme, M. (1992) Recent and future precipitation changes over the Nile Basin pp.187–201 in: *Climate fluctuations and water management* (eds) Biswas, A.K. and Abu Said, M.A., Butterworth-Heineman, Oxford, UK

Hulme, M. (2008) The conquering of climate: discourses of fear and their dissolution *Geographical Journal* 174, 5–16

Hulme, M. (2009) *Why we disagree about climate change: understanding controversy, inaction and opportunity* Cambridge University Press, Cambridge, UK

Hulme, M. (2010) Cosmopolitan climates: hybridity, foresight and meaning *Theory, Culture & Society* 27(2/3), 267–276

Hulme, M. (2011) Reducing the future to climate: a story of climate determinism and reductionism *Osiris* 26(1), 245–266

Hulme, M., Dessai, S., Lorenzoni, I. and Nelson, D. (2009) Unstable climates: exploring the statistical and social constructions of 'normal' climate' *Geoforum* 40(2), 197–206

Hulme, M., O'Neill, S.J. and Dessai, S. (2011) Is weather event attribution necessary for adaptation funding? *Science* 334, 764–765

Huntington, E. (1915) *Civilization and climate* Shoe String Press, Hamden CT

IPCC (1996a) *Climate Change 1995: the Science of Climate Change* Cambridge University Press, Cambridge, UK

IPCC (1996b) *Climate Change 1995: Impacts, Adaptations and Mitigation of Climate Change: Scientific-technical Analyses* Cambridge University Press, Cambridge, UK

IPCC (2007) *Climate change 2007: climate change impacts, adaptation and vulnerability* (eds) Parry, M., Canziani, O., Palutikof, J.P., van der Linden, P. and Hanson, C., WGII Contribution to the Fourth Assessment Report of the IPCC, Cambridge University Press, Cambridge, UK

IPCC (2012) *Managing the risks of extreme events and disasters to advance climate change adaptation* Cambridge University Press, Cambridge, UK

Jasanoff, S. (ed.) (2004) *States of knowledge: the co-production of science and the social order* Routledge, Abingdon, UK

Jasanoff, S. (2005) *Designs on nature: science and democracy in Europe and the United States* Princeton University Press, Princeton NJ

Jennings, T.L. (2009) Exploring the invisibility of local knowledge in decision-making: the Boscastle Harbour flood disaster pp.240–254 in: *Adapting to climate change: thresholds, values, governance* (eds) Adger, W.N., Lorenzoni, I. and O'Brien, K.L., Cambridge University Press, UK

Johns-Putra, A. and Trexler, A. (2013) *Anthropocene fictions: the novel in a time of climate change* University of Virginia Press, Charlestown VA

Jones, P.D., Raper, S.C.B. and Wigley, T.M.L. (1986) Global temperature variations, 1861–1984 *Nature* 322, 430–434

Jones, P.D., Wigley, T.M.L., Folland, C.K., Parker, D.E., Angell, J.K., Lebedeff, S. and Hansen, J.E. (1988a) Evidence for global warming in the past decade *Nature* 332, 790

Jones, P.D., Wigley, T.M.L., Folland, C.K. and Parker, D.E. (1988b) Spatial patterns in recent worldwide temperature trends *Climate Monitor*, 16, 175–185

Jones, P.D., Wigley, T.M.L. and Farmer, G. (1991) Marine and land temperature data sets: a comparison and look at recent trends pp.153–172 in: *Greenhouse-gas-induced climatic change: a critical appraisal of simulations and observations* (ed.) Schlesinger, M.E., Elsevier, Amsterdam, the Netherlands

Kahan, D. (2012) Why we are poles apart on climate change *Nature* 488, 255

Kahan, D., Jenkins-Smith, H. and Braman, D. (2011) Cultural cognition of scientific consensus *Journal of Risk Research* 14(2), 147–174

Kitcher, P. (2010) The climate change debates *Science* 328, 1230–1234

Lamb, H.H. (1982) *Climate, history and the modern world* Methuen, London, UK

Latour, B. (1993) *We have never been modern* (trans.) Harvard University Press, Harvard MA

Lee, N. and Motzkau, J. (2012) Varieties of biosocial imagination: reframing responses to climate change and antibiotic resistance *Science, Technology & Human Values* DOI: 10.1177/0162243912451498 (published on-line 16 July 2012)

Leggett, J. (ed.) (1990) *Global warming: the Greenpeace report* Oxford University Press, Oxford, UK

Leal Filho, W. and Manalos, E. (eds.) (2012) *English and climate change* Democritus University of Thrace, Thrace, Greece

Levene, M. (2010) Apocalypse as contemporary dialectic pp.59–80 in: *Future ethics: climate change and apocalyptic imagination* (ed.) Skrimshire, S., Continuum, London, UK

LICC Committee (eds) (1990) *Landscape ecological impact of climate change 1*, IOS, Amsterdam, the Netherlands

Lightman, B. (2007) *Victorian popularisers of science: designing nature for new audiences* University of Chicago Press, Chicago IL

Liverman, D. (2011) Book review symposium *Progress in Human Geography* 35(1), 132–138

Livingstone, D.N. (2003) *Putting science in its place: geographies of scientific knowledge* University of Chicago Press, Chicago IL

Lomborg, B. (ed.) (2004) *Global crises, global solutions* Cambridge University Press, Cambridge, UK

Lorenzoni, I., Nicholson-Cole, S. and Whitmarsh, L. (2007) Barriers perceived to engaging with climate change among the UK public and their policy implications *Global Environmental Change* 17(3/4), 445–459

Lovelock, J. (2006) *The revenge of Gaia: why the earth is fighting back - and how we can still save humanity* Penguin, London, UK

Lowe, P. and Phillipson, J. (2006) Reflexive interdisciplinary research: the making of a research programme on the rural economy and land use *Journal of Agricultural Economics*, 57, 165–184

Lowe, T., Brown, K., Dessai, S., de Franca Doria, M., Haynes, K. and Vincent, K. (2006) Does tomorrow ever come? Disaster narrative and public perceptions of climate change *Public Understanding of Science* 15(4), 435–457

Lozier, M.S. (2010) Deconstructing the conveyor belt *Science*, 328, 1507–1511

Lynas, M. (2004) *High tide: news from a warming world* Flamingo, London, UK

McKinnon, C. (2012) *Climate change and future justice: precaution, compensation and triage* Routledge, London, UK

Mahony, M. and Hulme, M. (2012) The colour of risk: an exploration of the IPCC's 'burning embers' diagram *Spontaneous Generation: A Journal for the History and Philosophy of Science* 6(1), 75–89

Maibach, E., Leiserowitz, A., Cobb, S., Shank, M., Cobb, K.M. and Gulledge, J. (2012) The legacy of Climategate: undermining or revitalizing climate science and policy? *WIREs Climate Change* 3(3), 289–295

Manley, G. (1952) *Climate and the British scene* Collins, London, UK

Meyer, S.M. (2006) *The end of the wild* Massachusetts Institute of Technology, Boston MA

Meyer, W.B. (2000) *Americans and their weather* Oxford University Press, Oxford, UK

Miller, C.A. (2004) Climate science and the making of a global political order pp.46–66 in, *States of knowledge: the co-production of science and the social order* (ed.) S. Jasanoff, Routledge, Abingdon, UK

Miller, C.A. (2007) Democratization, international knowledge institutions and global governance *Governance* 20(2), 325–357

Miller, C.A. and Edwards, P.N. (eds) (2001) *Changing the atmosphere: expert knowledge and environmental governance* MIT Press, Cambridge MA

Monbiot, G. (2006) *Heat: how to stop the planet burning* Allen & Lane, London, UK

Moon, T., Joughin, I., Smith, B. and Howat, I. (2012) 21st-century evolution of Greenland outlet glaciers velocities *Science* 336, 576–578

Moran, D. (ed.) (2011) *Climate change and national security: a country-level analysis* Georgetown University Press, Washington DC

Mouffe, C. (2005) *On the political* Routledge, Abingdon, UK

Newell, P. (2000) *Climate for change: non-state actors and the global politics of the greenhouse* (Hdbk) Cambridge University Press, Cambridge, UK

Nisbet, M.C., Hixon, M.A., Moore, K.D. and Nelson, M. (2010) Four cultures: new synergies for engaging society on climate change *Frontiers in Ecology and Environment* 8(6), 329–331

Nordhaus, W.D. and Boyer, J. (2000) *Warming the world: economic models of global warming* MIT Press, Cambridge MA

Norgaard, K.M. (2011) *Living in denial: Climate change, emotions and everyday life* MIT Press, Cambridge MA

Norgaard, R.B. (2002) Optimists, pessimists and science *BioScience*, 52(3), 287–292

O'Neill, S.J., Hulme, M., Turnpenny, J. and Screen, J.A. (2010) Disciplines, geography and gender in the framing of climate change *Bulletin of the American Meteorological Society* 91(8), 997–1002

O'Reilly, J., Oreskes, N. and Oppenheimer, M. (2012) The rapid disintegration of projections: the West Antarctic Ice Sheet and the IPCC *Social Studies of Science* 42(5), 709–731

Oreskes, N. and Conway, E. (2010) *Merchants of doubt: how a handful of scientists obscured the truth on issues from tobacco smoke to global warming* Bloomsbury Publishing, London, UK

Ostrom, E. (2010) Polycentric systems for coping with collective action and global environmental change *Global Environmental Change* 20(4), 550–557

Painter, J. and Ashe, T. (2012) Cross-national comparison of the presence of climate scepticism in the print media in six countries, 2007–2010 *Environmental Research Letters* 7, 044005

Parker, D.E., Folland, C.K. and Ward, M.N. (1988) Sea-surface temperature anomaly patterns and prediction of seasonal rainfall in the Sahel region of Africa pp.166–178 (Chapter 15) in: *Recent climatic change: a regional approach* (ed.) Gregory, S., Belhaven Press, London, UK

Parry, M.L., Carter, T.R. and Konijn, N.T. (1988) *The impact of climatic variations on agriculture: Volume 2, assessments in semi-arid regions* Kluwer Academic Publishers, Dordrecht, Germany

Parry, M.L., Hossell, J.E., Jones, P.J., Rehman, T., Tranter, R.B., Marsh, J.S., Rosenzweig, C., Fischer, G., Carson, I.G. and Bunce, R.G.H. (1996) Integrating global and regional analyses of the effects of climate change: a case study of land use in England and Wales *Climatic Change* 32, 185–198

Pettenger, M. (ed.) (2007) *The social construction of climate change* Ashgate Press, Aldershot, UK.

Petts, J., Owens, S. and Bulkley, H. (2008) Crossing boundaries: inter-disciplinarity in the context of urban environments *Geoforum* 39(2), 593–601

Philander, S.G. (ed.) (2008) *Encyclopedia of global warming and climate change* Sage, London, UK

Plass, G.N. (1956) The carbon dioxide theory of climate change *Tellus*, 8, 140–154

Pouillet, C.S.M. (1838) Mémoire sur la chaleur solaire, sur les pouvoirs rayonnants et absorbants de l'air atmosphérique, et sur la température de l'espace *Comptes rendus de l'Académie des Sciences*, 7, 24–65

Powell, R.C. (2007) Geographies of science: histories, localities, practises, futures *Progress in Human Geography* 31(3), 309–329

Prins, G. and Rayner, S. (2007) Time to ditch Kyoto *Nature* 449, 973–975

Prins, G., Galiana, I., Green, C., Grundmann, R., Hulme, M., Korhola, A., Laird, F., Nordhaus, T., Pielke Jr., R., Rayner, S., Sarewitz, D., Shellenberger, M., Stehr, N. and Tezuka, H. (2010) *The Hartwell Paper: a new direction for climate policy after the crash of 2009* London School of Economics, London, UK

Randalls, S. (2010) History of the 2°C climate target *WIREs Climate Change* 1(4), 598–605

Rayner, S. (2003) Domesticating nature: commentary on the anthropological study of weather and climate discourse pp.277–290 in, *Weather, climate, culture* (eds) Strauss, S. and Orlove, B., Berg, Oxford, UK

Rayner, S. and Malone, E.L. (eds) (1998) *Human choice and climate change, Volume 1 – the societal framework* Batelle Press, Columbus OH

Risbey, J.S. (2008) The new climate discourse: alarmist or alarming? *Global Environmental Change* 18(1), 26–37

Robinson, J. (2008) Being undisciplined: transgressions and intersections in academia and beyond *Futures* 40, 70–86

Royal Society (2012) *Science as an open enterprise* Royal Society, London, UK

Rudiak-Gould, P. (2013) *The sea also rises: facing climate change in the Marshall Islands* Routledge, Abingdon, UK

Russill, C. and Nyssa, Z. (2009) The tipping point trend in climate change communication *Global Environmental Change* 19(3), 336–344

Ryghaug, M. and Skjølsvold, T.M. (2010) The global warming of climate science: Climategate and the construction of scientific facts *International Studies in the Philosophy of Science* 24(3), 287–307

Sah, R. (1979) Priorities of developing countries in weather and climate *World Development* 7, 337–347

Sarewitz, D. (2004) How science makes environmental controversies worse *Environmental Science and Policy* 7, 385–403

Sarewitz, D. (2011) Does climate change knowledge really matter? *WIREs Climate Change* 2(4), 475–481

Sarewitz, D. and Pielke, R. Jr. (2000) Breaking the global-warming gridlock *The Atlantic Monthly* 286, 55–64

Saunier, R.E. and Meganck, R.A. (eds) (2007) *Dictionary and introduction to global environmental governance* Earthscan, London, UK

Schlesinger, M.E. and Mitchell, J.F.B. (1987) Climate model simulations of the equilibrium climatic response to increased dioxide *Review of Geophysics*, 15, 760–798

Schneider, S.H. (1977) Editorial for the first issue of Climatic Change *Climatic Change* 1, 3–4

Schneider, S.H. (1988) The greenhouse effect and the US summer of 1988: cause and effect or a media event? *Climatic Change* 13, 113–115

Schneider, S.H. (ed.) (1996) *Encyclopedia of climate and weather* (1st edn) Oxford University Press, Oxford, UK

Schneider, S.H. (2008) Geoengineering: could we or should we make it work? *Philosophical Transactions of the Royal Society (A)* 366, 3843–3862

Schneider, S.H. (ed.) (2011) *Encyclopedia of climate and weather* (2nd edn) Oxford University Press, Oxford, UK

Shapin, S. (1998) Placing the view from nowhere: historical and sociological problems in the location of science *Transactions of the Institute of British Geographers* 23, 5–12

Shapin, S. (2010) *Never pure: historical studies of science as if it was produced by people with bodies, situated in time, space, culture and society, and struggling for credibility and authority* The Johns Hopkins University Press, Baltimore MA

Sheffield, J., Wood, E.F. and Roderick, M.L. (2012) Little change in global drought over the past 60 years *Nature* 491, 435–438

Sherratt, T., Griffiths, T. and Robin, L. (eds) (2005) *A change in the weather: climate and culture in Australia* National Museum of Australia Press, Canberra, Australia

Shuckburgh, E., Robison, R. and Pidgeon, N. (2012) *Climate science, the public and the news media* Living With Environmental Change, Swindon, UK

Sindico, F. (2007) Climate change: a security (council) issue? *Carbon & Climate Law Review* 1(1), 29–34

Singer, F.S. (ed.) (2008) *Nature, not human activity, rules the climate: summary for policymakers of the report of the Nongovernmental International Panel on Climate Change* The Heartland Institute, Chicago IL

Skaggs, R.H. (2004) Climatology in American geography *Annals of the Association of American Geographers* 94(3), 446–457

Slocum, R. (2004) Polar bears and energy-efficient lightbulbs: strategies to bring climate change home *Environment & Planning D: Society and Space* 22, 413–438

Slovic, P. (2010) *The feeling of risk: new perspectives on risk perception* Earthscan, London, UK

Smith, D. and Vivekananda, J. (2007) A climate of conflict: the links between climate change, peace and war, *International Alert*, November

Solomon, S., Rosenlof, K.H., Portmann, R.W., Daniel, J.S., Davis, S.M., Sanford, T.J. and Plattner, G.-K. (2010) Contributions of stratospheric water vapour to decadal changes in the rate of global warming *Science* 327, 1219–1223

Speth, J.G. (2004) *Red sky at morning: America and the crisis of the global environment – a citizen's agenda for action* Yale University Press, Yale CT

Sterman, J.D. (2008) Risk communication on climate: mental models and mass balance *Science* 322, 532–533

Stern, N. (2007) *The Stern Review: The economics of climate change* Cambridge University Press, Cambridge, UK

Stoll-Kleeman, S., O'Riordan, T. and Jaeger, C.C. (2001) The psychology of denial concerning climate mitigation measures: evidence from Swiss focus groups *Global Environmental Change* 11, 107–117

Strauss, S. and Orlove, B. (eds) (2003) *Weather, climate, culture* Berg/Oxford International, Oxford, UK

Suk, J.E. (2012) Modelling consensus: science, politics and malaria at the IPCC Paper presented at the 4S/EASST Conference, Copenhagen, October 2012

Thomas, D.S.G. and Goudie, A.S. (eds) (2000) *Dictionary of physical geography* Wiley-Blackwell, Chichester, UK

Thornes, J.E. (2008) Cultural climatology and the representation of sky, atmosphere, weather and climate in selected art works of Constable, Monet and Eliasson *Geoforum* 39(2), 570–580

Thornes, J.E. and Randalls, S. (2007) Commodifying the atmosphere: 'pennies from heaven'? *Geografiska Annaler*, 89A(4) 273–285

Tol, R.S.J. (2007) Europe's long-term policy goal: a critical evaluation *Energy Policy* 35, 424–432

Trexler, A. and Johns-Putra, A. (2011) Climate change in literature and literary criticism *WIREs Climate Change* 2(2), 185–200

Turner, F.M. (1981) John Tyndall and Victorian scientific naturalism pp.169–180 in: *John Tyndall: essays on a natural philosopher* (eds) Brock, W.H., McMillan, N.D. and Mollan, R.C., Royal Dublin Society, Dublin, Ireland

Tyndall, J. (1855–1872) *Journals of John Tyndall Vol.3, 1855–1872* Journal 8a, pp.36–55 (May 1859) & pp.226–230 (February 1861) Unpublished Tyndall Collection, Royal Institution, London, UK

Tyndall, J. (1859a) On the transmission of heat of different qualities through gases of different kinds *Proceedings of the Royal Institution* 3, 155–158

Tyndall, J. (1859b) Letter to Clausius 1 June 1859 130/B10.7 Cox S., Packet 7 Unpublished Tyndall Collection, Royal Institution, London, UK

Tyndall, J. (1860) *The glaciers of the Alps* Murray, London, UK

Tyndall, J. (1861) On the absorption and radiation of heat by gases and vapours *Philosophical Magazine* 4th series, 22, 169–194 & 273–285

Tyndall, J. (1863) On the passage of radiant heat through dry and humid air *Philosophical Magazine* 4th series, 26, 44–54.

UNEP (2007) Sudan post-conflict environment assessment UNEP, Nairobi, Kenya

UNEP (2010) Busan Outcome, 7–11 June 2010 UNEP, Nairobi, Kenya

UNFCCC (1992) Article 2, the *United Nations Framework Convention on Climate Change* United Nations, New York

UNFCCC (2009) Decision 2/CP.15, Copenhagen Accord, December 7–19 2009, http://unfccc.int/resource/docs/2009/cop15/eng/l07.pdf (accessed 31 December 2012)

Van der Sluijs, J., van Eijndhoven, J. Shackley, S. and Wynne, B. (1998) Anchoring devices in science for policy: the case of consensus around the climate sensitivity *Social Studies of Science* 28(2), 291–323

Veldman, R.G., Szasz, A. and Haluza-DeLay, R. (eds) (2013) *How the world's religions are responding to climate change: social scientific investigations* Routledge, Abingdon, UK

Von Neumann, J. (1955) Can we survive technology? *Fortune* 91(6), 32–47

Walker, G. and King, D. (2008) *The hot topic: how to tackle climate change and still keep the lights on* Bloomsbury Press, London, UK

Weaver, C.P., Lempert, R.J., Brown, C., Hall, J.A., Revell, D. and Sarewitz, D. (2013) Improving the contribution of climate model information to decision-making: the value and demands of robust decision frameworks *WIREs Climate Change* 4(1), 39–60

Whatmore, S.J. (2009) Mapping knowledge controversies: science, democracy and the redistribution of expertise *Progress in Human Geography* 33(5), 587–598

Wilbanks, T.J. and Kates, R.W. (1999) Global change in local places: how scale matters *Climatic Change* 43(3), 601–628

Wilkins, L. (1993) Between facts and values: print media coverage of the greenhouse effect, 1987–1990 *Public Understanding of Science* 2, 71–84

Wilkinson, K.K. (2012) *Between God and green: how evangelicals are cultivating a middle ground on climate change* Oxford University Press, New York NY

Woods Hole Research Centre (1989) Conference Statement from the International Conference on Global Warming and Climate Change: Perspectives from Developing Countries, New Delhi, 21–23 February, Woods Hole, MA

Yearley, S. (2006) How many 'ends' of nature: making sociological and phenomenological sense of the end of nature *Nature and Culture* 1(1), 10–21

Yusoff, K. and Gabrys, J. (2011) Climate change and the imagination *WIREs Climate Change* 2(4), 516–534

Zhang, D.D., Zhang, J., Lee, H.F. and Ye, H. (2007) Climate change and war frequency in eastern China over the last millennium *Human Ecology* 34, 403–414

Index

The suffix 'n' indicates a footnote on the indicated page.